Cod

Mark Kurlansky is the author of *A Continent of Islands: Searching for the Caribbean Destiny*, *A Chosen Few: The Resurrection of European Jewry* and *The Basque History of the World*. During the past twenty years he has spent a great deal of time in the Caribbean, including seven years as the *Chicago Tribune*'s Caribbean correspondent. His latest book is a collection of short stories, *The White Man in the Tree*. He lives in New York City.

ALSO BY MARK KURLANSKY

Non-fiction

A Continent of Islands: Searching for the Caribbean Destiny
A Chosen Few: The Resurrection of European Jewry
The Basque History of the World

Fiction

The White Man in the Tree

Mark Kurlansky

COD

A Biography of the Fish That Changed the World

VINTAGE BOOKS
London

Published by Vintage 1999

22

First published in the United States of America in 1997
by Walker Publishing Company, Inc.
First published in Great Britain by Jonathan Cape in 1998

Vintage
Random House, 20 Vauxhall Bridge Road, London SW1V 2SA
www.vintage-books.co.uk

The Random House Group Limited Reg. no. 954009

A CIP catalogue record for this book
is available from the British Library

ISBN 9780099268703

Book Design by M.J. Di Massi

Printed and bound by CPI Group (UK) Ltd, Croydon, CR0 4YY

THE QUESTION OF QUESTIONS FOR MANKIND—THE PROBLEM WHICH UNDERLIES ALL OTHERS, AND IS MORE DEEPLY INTERESTING THAN ANY OTHER—IS THE ASCERTAINMENT OF THE PLACE WHICH MAN OCCUPIES IN NATURE AND OF HIS RELATIONS TO THE UNIVERSE OF THINGS.

—H. Thomas Henry Huxley,
Man's Place in Nature

SO THE FIRST BIOLOGICAL LESSON OF HISTORY IS THAT LIFE IS COMPETITION. COMPETITION IS NOT ONLY THE LIFE OF TRADE, IT IS THE TRADE OF LIFE—PEACEFUL WHEN FOOD ABOUNDS, VIOLENT WHEN THE MOUTHS OUTRUN THE FOOD. ANIMALS EAT ONE ANOTHER WITHOUT QUALM; CIVILIZED MEN CONSUME ONE ANOTHER BY DUE PROCESS OF LAW."

—Will and Ariel Durant,
The Lessons of History

Contents

PART THREE: THE LAST HUNTERS

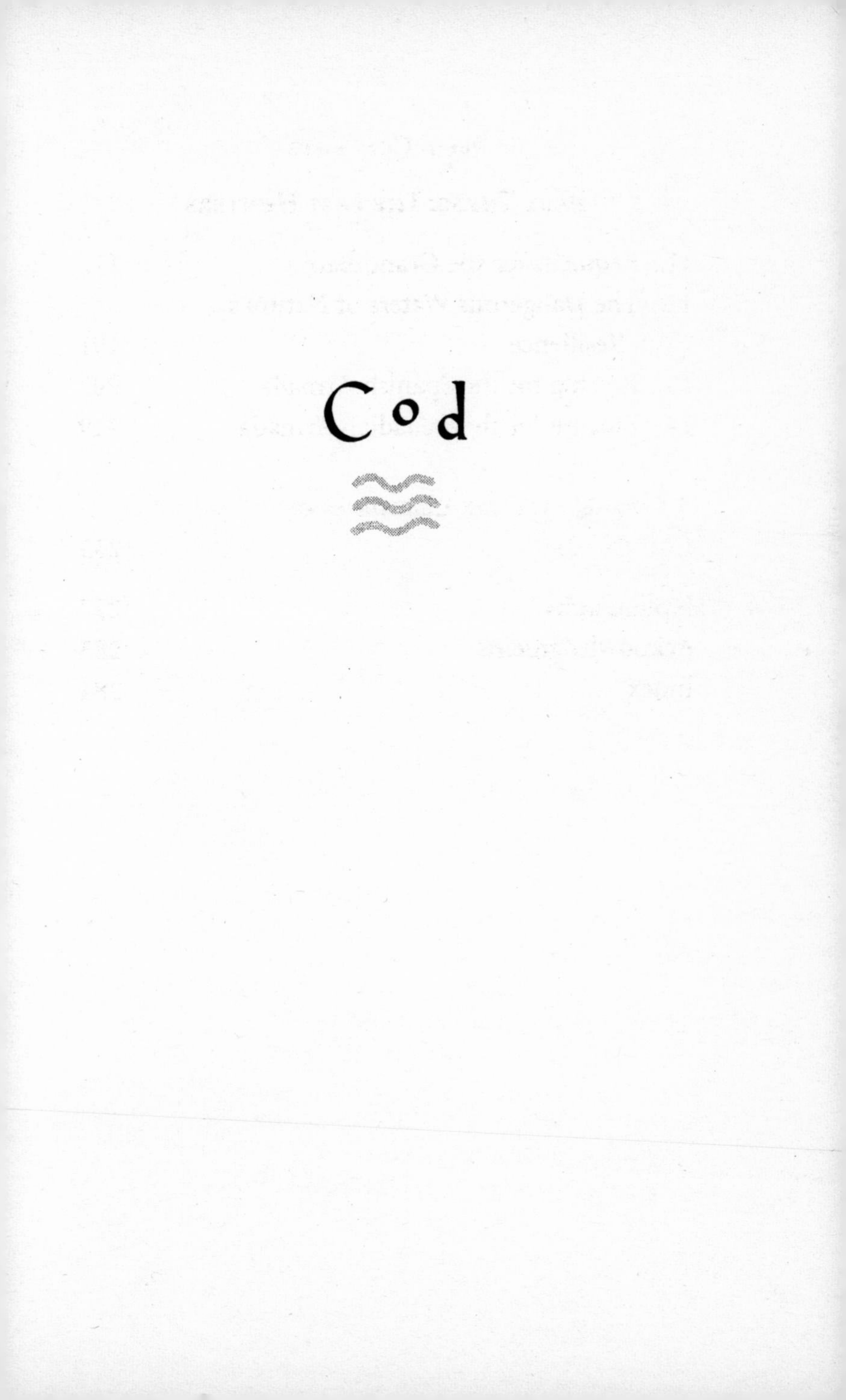

Cod

Prologue: Sentry on the Headlands (So Close to Ireland)

THE HERRING ARE NOT IN THE TIDES AS THEY
WERE OF OLD;
MY SORROW FOR MANY A CREAK GAVE THE CREEL
IN THE CART
THAT CARRIED THE TAKE TO SLIGO TOWN TO BE SOLD,
WHEN I WAS A BOY WITH NEVER A CRACK IN MY HEART.

—William Butler Yeats,
"The Meditation of the Old Fisherman"

These are the fishermen who stand sentry over the cod stocks off the headlands of North America, the fishermen who went to sea but forgot their pencil.

Sam Lee, dressed in black rubber boots and a red flotation jacket made even brighter by its newness, drives his late-model pickup truck through the last murk of night, down to the wharves that stretch out to where the water is deep enough for a shallow fishing skiff. The warehouses, meeting halls, and tackle shops are all built out above the shallow water on stilts. This has freed up the narrow strip of flat land where the steep little mountains stop just before the water's edge. The level area had once been needed to spread out thousands of splayed and salted cod for drying in the open air.

The salting had stopped almost thirty years before, but Petty Harbour still looks like a crowded little port, its few commercial buildings crunched in along the water, while houses scatter up onto the beginnings of the slopes.

At the wharves, Sam meets up with Leonard Stack and Bernard Chafe carrying flashlights and joking about Sam's new jacket, shielding their eyes from its shocking brilliance. Grumbling about fishery politics, about last night's television talk of reopening groundfishing to the public on a limited basis, they climb down into Leonard's thirty-two-foot, open-decked trap skiff.

Asked if he could really float with that jacket, Sam answers, "I don't want to find out!" And that is all they say about the black water a few feet away on either side of them as the boat heads out in the first violet light of early-autumn morning. Cod like the water this time of year because they think it is warm. But forty-five degrees Fahrenheit is a cod's idea of warm, and the gunwales around the edge of a trap skiff are only inches high. This same day, in another community, the bodies of two fishermen who fell overboard are found. This isn't something fishermen talk about.

They head out to sea. Sam, a small, dark-haired man, with a touch of rose on his clean-shaven cheeks, is stuffed into his scarlet flotation jacket. Leonard is in the little pilothouse, while Bernard, in his flame orange overalls, stands with Sam on the open deck looking contemplatively at a flat sea of dark, polished facets. The light is beginning to warm a clear sky. Once the sun is up, the only clouds are cotton candy fog stuck between the rocky, still-green hills of the September coastline.

They find their fishing grounds by land markings. When a brown rock is aligned over the church steeple, when certain houses first come into view, or when they first sight the white spot on a rock that they call "the Madame" because in their imagination it looks like a skirt and a bonnet, they are ready to drop anchor and begin fishing.

Only today, having forgotten a pencil, they head over to the other boat where the three-man crew is already hauling cod with handlines. After a few jokes about the size of this sorry young catch, someone tosses over a pencil. They are ready to fish.

These men are part of the Sentinel Fishery, now the only legal cod fishery in Newfoundland. In July 1992, the Canadian government closed Newfoundland waters, the Grand Banks, and most of the Gulf of St. Lawrence to groundfishing. Groundfish, of which the most sought after is cod, are those that live in the bottom layer of the ocean's water. By the time the moratorium was announced, the fishermen of Petty Harbour, seeing the rapid decline of their once prolific catch, had been demanding it for years. They had claimed, and it is now acknowledged, that the offshore trawlers were taking nearly every last cod. In the 1980s, government scientists had ignored the cry of inshore fishermen that the cod were disappearing. This deafness proved costly.

Now two Petty Harbour boats are participating in the Sentinel Fishery, a program meant to get scientists and fishermen working together. A few fishermen in each community are sentries, measuring the progress of the cod stock by catching fish and reporting their findings to government scientists. The men on Leonard's boat are tagging and releasing as many fish as they can catch. At the same time, the fishermen on the other boat are supposed to catch exactly 100 fish, open them up to see if they are male or female, and remove a tiny bone from the head, the otolith, which helps the cod keep its balance. The rings of the otolith tell the cod's age.

Tomorrow, or the next good day with a calm sea, the two boats will switch jobs. There is no point in braving bad weather. The fishermen earn only a modest rent on their boats for this work but are glad to have it, because it gives them something to do besides collecting their unemployment compensation, renowned in Maritime Canada as "the package." They also like doing it because there is constant pressure to reopen fishing. This week the debate is on an idea to let everyone fish a few cod "just for food." The Sentinel fishermen are proving with their scant, undersized, and underaged catch that there are still not enough cod to allow any fishing at all.

"This is it. We are out on the headlands," Sam frequently reminds people. Petty Harbour fishermen are proud of the fact that they live in the most easterly fishing community in North America—the first of three things for which Petty Harbour is famous. Their little village, along with St. John's in the next cove and the rocky point between them, is the site closest to the part of the North Atlantic fabled throughout this millennium as the cod grounds.

Being on the eastern headlands also means that it is the North American town closest to Ireland, and this is the second thing for which the town is famous. Although the name Petty Harbour comes from the French *petit,* the people here are Irish. Fifth-generation Newfoundlanders speak with the musical brogue of southern Ireland. While this accent is heard up and down the Newfoundland coast, Petty Harbour is a microcosm of Ireland—Ireland upside down. The village, with its population of almost 1,000, was built on the mouth of a small river. On the north side live the Catholics. On the south are the Protestants. The little bridge was a border, and the people on either side never mixed. Sam, Leonard, and Bernard are all Catholic. But, growing up in the late 1950s and early 1960s, they were the first generation of children to cross the bridge when playing. Sam married a Protestant. So did Bernard, who, now forty-one, is five years younger than Sam. The town's sole social conflict faded—only to be replaced by new ones as the cod disappeared.

According to Sam, it is more than the cod that are gone. He looks toward the horizon and says, "Not a whale, nothing." For years, he has not seen herring or capelin, which the humpback whale chase. The squid, too, seems to have vanished. Petty Harbour fishermen used to spend an hour jigging the harbor for squid to use as bait. This morning, they are using squid they bought frozen.

In summer, before their disappearance, the cod would come so close to shore that fishermen could catch them in traps, ingenious devices invented in Labrador in the nineteeth century. A wall of twine net was anchored to the shore, and cod swimming from either side followed the wall and found themselves in a large, twine, underwater room, which they could easily leave. But most didn't. The unbaited traps were left out in July and August and hauled up twice a day. Thousands of cod used to swim into these traps along the rocky coast in the summer. At the time of the moratorium, the 125 fishermen of Petty Harbour were setting seventy-five traps along the deep inlet that marked Petty Harbour waters.

Then, in September, when the cod started moving farther offshore, the handlining season would begin. Handline fishing dates back to the iron age. A hook is baited, and a four-ounce lead weight drops it to the bottom on heavy line. In Petty Harbour's grounds, the men fish at a depth between fifteen and thirty fathoms. The fisherman loops the line around his hand and when he feels a tug, he yanks hard to set the hook in the fish's mouth. He must yank the line and start pulling it in with one continuous motion, because any slack will enable the fish to wriggle free. But few escape these fishermen.

Once the hook is well set, the cod doesn't fight and it is simply a matter of hauling up the weight. The skill is all in the first moments; the rest is labor. The fishermen rapidly haul in some 180 feet of line by moving their two index fingers in broad circular motions. In the old days, they would each have had a line out: two on the side of the boat where the tide runs and one on the opposite side. The open deck and low gunwales might be dangerous in a rough sea, but they make it easy to land fish. The three men would have hauled up fish weighing from eight to thirty pounds or more, one after another, without a break, for the rest of the day until the deck and both three-foot-deep holds had no more room. Each boat would have returned with between 2,000 and 3,000 pounds of cod. Fifty or more boats from Petty Harbour would have all been out there with two- or three-man crews, hauling fish and shouting jokes from boat to boat.

The third thing for which Petty Harbour is famous: The community has banned the mass-fishing techniques of longlining and gillnetting since the late 1940s. Since the moratorium, environmentalists have singled out Petty Harbour for having taken this step decades before anyone else in Newfoundland was talking about conservation. In 1995, the Sierra Club, the conservation group, noted in its magazine: "More than a generation ago, Petty Harbour fishermen outlawed destructive practices like trawling and gillnetting. Petty Harbour allows only conservation-oriented fishing gear—old-fashioned handlines . . . and traps."

But, in truth, the ban was implemented because with 125 fishermen working the opening of the same cove, there simply would not have been enough space for such practices. "Nowadays, everyone tries to say that it was for conservation," says Sam. "There was no such thing as conservation. For God's sake, there were enough fish to walk on. It was because there wasn't enough room."

Newfoundland's inshore fishermen fish only the waters of their own cove. If a Petty Harbour boat wanted to work beyond the last point of rock in Petty Harbour's inlet, he would ask the St. John's fishermen in the neighboring cove for permission. That was back in the days of civility, before the moratorium, when there were supposed to be enough fish for everyone, and religion was the only bone to fight over.

Since the moratorium was declared, civility has been scarcer than cod. Six Petty Harbour boats even went gill-netting in plain view, and it took two years of legal action and political pressure to stop them.

Commercially, Sam, Bernard, and Leonard do not fish together. Sam used to work with his brother. Bernard's partner of twenty years never got a groundfishing license when they were easy to get. He hadn't needed one. Now, if groundfishing ever opens up again, there will be a strict fish-per-license quota and no new licenses will be available. Bernard will have to share his quota with his partner, and it probably will not be a big enough catch for two. "And I'm supposed to tell the man I've been fishing with all these years, 'Sorry, I have to team up with someone with a groundfishing license.' They want to make people leave fishing. But what else is there?"

"It used to be a nice place to live," says Sam, "but it's not anymore."

"It's unbelievable," says Bernard, "the way a few years ago everybody just did what they did, and they didn't worry about anyone else. Now no one wants to see anyone make a dollar that they're not making. Everybody is watching everybody else. I don't think you can fart in the community without someone complaining."

But on this perfect Newfoundland September morning with a warming sun and a flat sea, these men of the Sentinel Fishery are in a good mood, doing the only thing they have ever wanted to do, going out on the water with their childhood friends to haul up fish.

The catch is a disaster.

Newfoundland and Labrador cod, the so-called northern stock, are pretty fish with amber leopard spots on an olive green back, a white belly, and the long white, streamlining stripe between the belly and the spotted back. They are far prettier than the Icelandic stock, with its yellow on brown. The fishermen measure each cod as it is hauled in and find that the length ranges from forty-five to fifty-five centimeters (twenty inches or so), which means they are two- or three-year-old codlings born since the moratorium—not even old enough to reproduce. When Leonard finally hauls up a cod of seventy-five centimeters, probably seven years old, a typical catch ten years ago, they all joke, "Oh, my God, get the gaff! Give him a hand!"

In their lilting brogues, they joke about the fact that they are not real fishermen anymore. The little boat hits a slight swell sideways, and as it rolls Sam whines, "Ohhh, I think I'm going to be seasick." The others laugh.

They are good at hauling up fish. But this is something different. Instead of throwing the cod on the deck and quickly baiting and recasting for the next, they have to gently remove the hook and try not to hurt the animal. Then they lay it out on a board and measure it in centimeters. A tool with a trigger mechanism is used to insert an inch-long needle in the meaty part next to the forward dorsal fin and snap into place a plastic thread with a numbered tag on the end. This they are not very good at.

Sam unhooks a fish, and it jerks out of his hands and crashes to the deck. "Oh, sorry," he says to the cod in the same tender little voice he uses at home with his aging beagle. The tagging gun is not working well, so Sam takes it apart and rebuilds it. Taking things apart and fixing them is part of a fisherman's skills. But the gun still doesn't work well. Sometimes they have to stick a fish three or four times to get a tag in. This is proof of what a tough survivor the cod is. A salmon would never survive this handling. But when they finally drop the codfish in the sea slowly, head first, to revive it, it instantly swims for its home on the bottom. To a cod, ocean floors mean safety. That is why they were rendered commercially extinct by bottom draggers.

Trying to insert the tag in one cod, the men stick and poke it so many times that it dies. That makes no one sad, because they are hungry. Bernard kneels over a portable Sterno stove at the stern. He uses his thick fishing knife to dice fatback and salt beef and peel and slice potatoes. He soaks pieces of hardtack and sautés it all in the pork fat with some sliced onion. Then he fillets the cod in four knife strokes per side, skins the fillets with two more, and before throwing the carcass over, opens it up, sees it is a female, and removes the roe. Holding it by a gill over the gunwale, he makes two quick cuts and rips out the throat piece, "the cod tongue," before dropping the body in the sea.

As Bernard stirs his pot, Sam records tag numbers and fish lengths with his pencil, while at the bow Leonard silently hauls up one young cod after another with his fast-moving gloved forefingers. "Leonard's having all the fun," Bernard says in mock grumpiness.

Bernard dumps the food on a big baking sheet, which they put on a plank across one of the holds, and they stand in the hold where the catch should have been and with plastic forks start eating toward the center. The dish, called Fishermen's Brewis, is monochromatic, with off-white pork fat and off-white potatoes and occasional darker pieces of salt beef. What stands out is the stark whiteness of the thick flakes of fresh cod. This is the meal they grew up on, and, as often happens when old friends are eating their childhood food, they start reminiscing.

There were no sports for these men to talk about, no high school teams; they aren't even hockey fans. As children, they went fishing with their fathers every morning just before daybreak. They would come to shore mid-day and go to school—until the first black cloud passed overhead and they had to run down to the harbor, to the racks, called fish flakes, where the salt cod were drying, and turn them over skin side up so they would be protected if it rained.

Instead of sports, they talk about fishing, about how cold it used to be. It is not that the weather has changed. But back then, there had been no lightweight microfibers to hold in body heat, nothing to help the fingers reeling in line dripping with icy water. All this in a season with little sun, or even daylight, for warmth. The fishing was good into January, but when, in 1957, unemployment compensation was made available for fishermen after December 15, that became the last fishing day until spring. Years later the date was moved to November 15.

But they remember fishing into the winter. "Christ," says Bernard, "out there handlining in the snow. You'd come in numb. We didn't have these modern clothes. Just wool. Or if they had them, we didn't know."

"No," said Leonard, "they weren't there."

"Christ, it was cold."

"We didn't have any choices."

"Couldn't even put this much salt beef in."

The conversation turns to a favorite Newfoundland topic, how unhealthy their diet is. Traditional Newfoundland food is based on pork fat. Everything is cooked in it and then seasoned with scrunchions—rendered, diced fatback.

"Good for the arteries," Bernard says with a laugh. "You know what my brother says. You put something in front of him, and he always asks, 'Is it good for you?' If you say 'Yes,' he says, 'Then I don't want it.' "

They finish eating—Sam and Bernard share the roe, and Leonard eats the tongue—and head back to harbor. Only forty fish have been tagged, and the biggest is just seventy-six centimeters (thirty inches). Ten years ago, this record fish would have been barely the average size. Only three of the forty are large enough to be capable of spawning.

The men in the other boat worked three lines and caught their 100 fish with a total weight of 375 pounds. This means the average is less than four pounds at the time of year when Petty Harbour used to get some of its biggest catches—boats with 300 fish having a total weight of 3,000 pounds.

They set aside the parts for the scientists and divide the rest of the fish into bags containing about ten pounds of fish each. A ten-pound bag should have been one cod, but most bags have two or three. When the two boats come into the harbor, some fifty people, mostly from other towns, are already waiting in a polite line.

This is Canada. These people have jobs or are on public assistance, mostly the latter these days. They are not hungry but simply yearning for a taste of their local dish. The big fish companies, the ones that owned bottom draggers that had cleaned out the last of the cod before the moratorium, now import frozen cod from Iceland, Russia, and Norway. But these people are accustomed to fresh, white, flaky cod "with the nerves still tingling," as one fisherman's daughter put it. Sam had once sent a shipment to New Orleans, and the chef had complained that it was too fresh and the meat did not hold together well. Only fishing communities know what real fresh cod, with thick white flakes that come apart, tastes like.

Even limiting the cod to ten pounds a person, there

is not enough. A few people are turned away, and one of them asks one of the fishermen, "Where are they taking the rest of the fish?"

The problem with the people in Petty Harbour, out here on the headlands of North America, is that they are at the wrong end of a 1,000-year fishing spree.

part one

A Fish Tale

. . . SALT COD, SPREADING ITSELF BEFORE THE DRAB, HEFTY SHOP KEEPERS, MAKING THEM DREAM OF DEPARTURE, OF TRAVEL.

—Émile Zola,
"The Belly of Paris," 1873

1: The Race to Codlandia

HE SAID IT MUST BE FRIDAY, THE DAY HE COULD NOT SELL ANYTHING EXCEPT SERVINGS OF A FISH KNOWN IN CASTILE AS POLLOCK OR IN ANDALUSIA AS SALT COD.

—Miguel de Cervantes,
Don Quixote, 1605–1616

A medieval fisherman is said to have hauled up a three-foot-long cod, which was common enough at the time. And the fact that the cod could talk was not especially surprising. But what *was* astonishing was that it spoke an unknown language. It spoke Basque.

This Basque folktale shows not only the Basque attachment to their orphan language, indecipherable to the rest of the world, but also their tie to the Atlantic cod,

Gadus morhua, a fish that has never been found in Basque or even Spanish waters.

The Basques are enigmatic. They have lived in what is now the northwest corner of Spain and a nick of the French southwest for longer than history records, and not only is the origin of their language unknown, but the origin of the people themselves remains a mystery also. According to one theory, these rosy-cheeked, dark-haired, long-nosed people were the original Iberians, driven by invaders to this mountainous corner between the Pyrenees, the Cantabrian Sierra, and the Bay of Biscay. Or they may be indigenous to this area.

They graze sheep on impossibly steep, green slopes of mountains that are thrilling in their rare, rugged beauty. They sing their own songs and write their own literature in their own language, Euskera. Possibly Europe's oldest living language, Euskera is one of only four European languages—along with Estonian, Finnish, and Hungarian—not in the Indo-European family. They also have their own sports, most notably jai alai, and even their own hat, the Basque beret, which is bigger than any other beret.

Though their lands currently reside in three provinces of France and four of Spain, Basques have always insisted that they have a country, and they call it Euskadi. All the powerful peoples around them—the Celts and Romans, the royal houses of Aquitaine, Navarra, Aragon, and Castile; later Spanish and French monarchies, dictatorships, and republics—have tried to subdue and assimilate them, and all have failed. In the 1960s, at a time when their ancient language was only whispered, having been outlawed by the dictator Francisco Franco, they secretly modernized it to broaden its usage, and today, with only 800,000 Basque speakers in the world, almost 1,000 titles a year are published in Euskera, nearly a third by Basque writers and the rest translations.

"*Nire aitaren etxea / defendituko dut. / Otsoen kontra*" (I will defend / the house of my father. / Against the wolves) are the opening lines of a famous poem in modern Euskera by Gabriel Aresti, one of the fathers of the modernized tongue. Basques have been able to maintain this stubborn independence, despite repression and wars, because they have managed to preserve a strong economy throughout the centuries. Not only are Basques shepherds, but they are also a seafaring people, noted for their successes in commerce. During the Middle Ages, when Europeans ate great quantities of whale meat, the Basques traveled to distant unknown waters and brought back whale. They were able to travel such distances because they had found huge schools of cod and salted their catch, giving them a nutritious food supply that would not spoil on long voyages.

Basques were not the first to cure cod. Centuries earlier, the Vikings had traveled from Norway to Iceland to Greenland to Canada, and it is not a coincidence that this is the exact range of the Atlantic cod. In the tenth century, Thorwald and his wayward son, Eirik the Red, having been thrown out of Norway for murder, traveled to Iceland, where they killed more people and were again expelled. About the year 985, they put to sea from the black lava shore of Iceland with a small crew on a little open ship. Even in midsummer, when the days are almost without nightfall, the sea there is gray and kicks up whitecaps. But with sails and oars, the small band made it to a land of glaciers and rocks, where the water was treacherous with icebergs that glowed robin's-egg blue. In the spring and summer, chunks broke off the glaciers, crashed into the sea with a sound like thunder that echoed in the fjords, and sent out huge waves. Eirik, hoping to colonize this land, tried to enhance its appeal by naming it Greenland.

Almost 1,000 years later, New England whalers would sing: "Oh, Greenland is a barren place / a place that bears no green / Where there's ice and snow / and the whale fishes blow / But daylight's seldom seen."

Eirik colonized this inhospitable land and then tried to push on to new discoveries. But he injured his foot and had to be left behind. His son, Leifur, later known as Leif Eiriksson, sailed on to a place he called Stoneland, which was probably the rocky, barren Labrador coast. "I saw not one cartload of earth, though I landed many places," Jacques Cartier would write of this coast six centuries later. From there, Leif's men turned south to "Woodland" and then "Vineland." The identity of these places is not certain. Woodland could have been Newfoundland, Nova Scotia, or Maine, all three of which are wooded. But in Vineland they found wild grapes, which no one else has discovered in any of these places.

The remains of a Viking camp have been found in Newfoundland. It is perhaps in that gentler land that the

Vikings were greeted by inhabitants they found so violent and hostile that they deemed settlement impossible, a striking assessment to come from a people who had been regularly banished for the habit of murdering people. More than 500 years later the Beothuk tribe of Newfoundland would prevent John Cabot from exploring beyond crossbow range of his ship. The Beothuk apparently did not misjudge Europeans, since soon after Cabot, they were enslaved by the Portuguese, driven inland, hunted by the French and English, and exterminated in a matter of decades.

How did the Vikings survive in greenless Greenland and earthless Stoneland? How did they have enough provisions to push on to Woodland and Vineland, where they dared not go inland to gather food, and yet they still had enough food to get back? What did these Norsemen eat on the five expeditions to America between 985 and 1011 that have been recorded in the Icelandic sagas? They were able to travel to all these distant, barren shores because they had learned to preserve codfish by hanging it in the frosty winter air until it lost four-fifths of its weight and became a durable woodlike plank. They could break off pieces and chew them, eating it like hardtack. Even earlier than Eirik's day, in the ninth century, Norsemen had already established plants for processing dried cod in Iceland and Norway and were trading the surplus in northern Europe.

The Basques, unlike the Vikings, had salt, and because fish that was salted before drying lasted longer, the

Basques could travel even farther than the Vikings. They had another advantage: The more durable a product, the easier it is to trade. By the year 1000, the Basques had greatly expanded the cod markets to a truly international trade that reached far from the cod's northern habitat.

In the Mediterranean world, where there were not only salt deposits but a strong enough sun to dry sea salt, salting to preserve food was not a new idea. In preclassical times, Egyptians and Romans had salted fish and developed a thriving trade. Salted meats were popular, and Roman Gaul had been famous for salted and smoked hams. Before they turned to cod, the Basques had sometimes salted whale meat; salt whale was found to be good with peas, and the most prized part of the whale, the tongue, was also often salted.

Until the twentieth-century refrigerator, spoiled food had been a chronic curse and severely limited trade in many products, especially fish. When the Basque whalers applied to cod the salting techniques they were using on whale, they discovered a particularly good marriage because the cod is virtually without fat, and so if salted and dried well, would rarely spoil. It would outlast whale, which is red meat, and it would outlast herring, a fatty fish that became a popular salted item of the northern countries in the Middle Ages.

Even dried salted cod will turn if kept long enough in hot humid weather. But for the Middle Ages it was remarkably long-lasting—a miracle comparable to the discovery of the fast-freezing process in the twentieth century, which also debuted with cod. Not only did cod

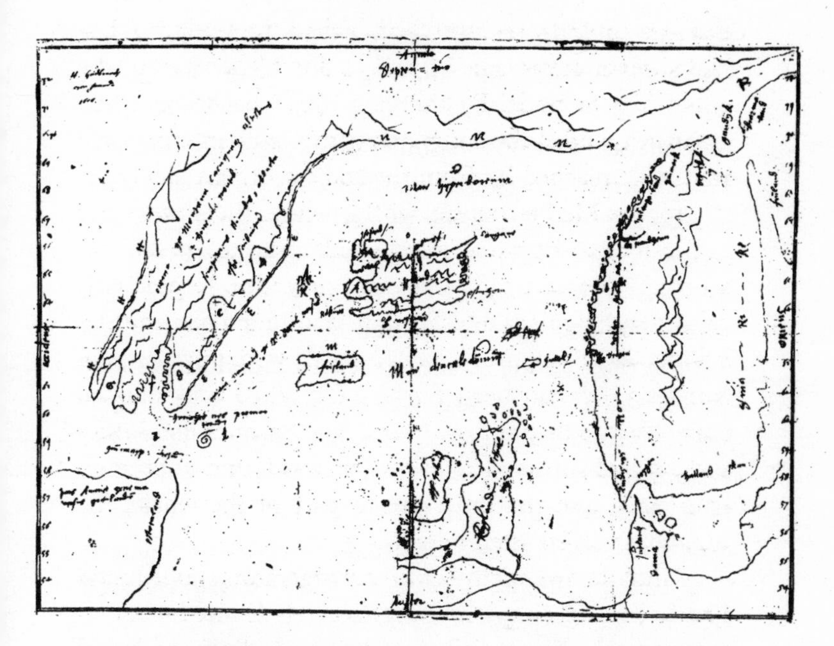

In 1606, Gudbrandur Thorláksson, an Icelandic bishop, made this line drawing of the North Atlantic in which Greenland is represented in the shape of a dragon with a fierce, toothy mouth. Modern maps show that this is not at all the shape of Greenland, but it is exactly what it looks like from the southern fjords, which cut jagged gashes miles deep into the high mountains. (Royal Library, Copenhagen)

last longer than other salted fish, but it tasted better too. Once dried or salted—or both—and then properly restored through soaking, this fish presents a flaky flesh that to many tastes, even in the modern age of refrigeration, is far superior to the bland white meat of fresh cod.

For the poor who could rarely afford fresh fish, it was cheap, high-quality nutrition.

Catholicism gave the Basques their great opportunity. The medieval church imposed fast days on which sexual intercourse and the eating of flesh were forbidden, but eating "cold" foods was permitted. Because fish came from water, it was deemed cold, as were waterfowl and whale, but meat was considered hot food. The Basques were already selling whale meat to Catholics on "lean days," which, since Friday was the day of Christ's crucifixion, included all Fridays, the forty days of Lent, and various other days of note on the religious calendar. In total, meat was forbidden for almost half the days of the year, and those lean days eventually became salt cod days. Cod became almost a religious icon—a mythological crusader for Christian observance.

The Basques were getting richer every Friday. But where was all this cod coming from? The Basques, who had never even said where they came from, kept their secret. By the fifteenth century, this was no longer easy to do, because cod had become widely recognized as a highly profitable commodity and commercial interests around Europe were looking for new cod grounds. There were cod off of Iceland and in the North Sea, but the Scandinavians, who had been fishing cod in those waters for thousands of years, had not seen the Basques. The British, who had been fishing for cod well offshore since Roman times, did not run across Basque fishermen even in the fourteenth century, when British fishermen began venturing up to Icelandic waters. The

Bench ends from St. Nicolas' Chapel in a town by the North Sea, King's Lynn, Norfolk, England, carved circa 1415, depict the cod fishery. (Victoria and Albert Museum, London)

Bretons, who tried to follow the Basques, began talking of a land across the sea.

In the 1480s, a conflict was brewing between Bristol merchants and the Hanseatic League. The league had been formed in thirteenth-century Lübeck to regulate trade and stand up for the interests of the merchant class in northern German towns. *Hanse* means "fellowship" in Middle High German. This fellowship organized town by town and spread throughout northern Europe, including London. By controlling the mouths of all the major rivers that ran north from central Europe, from the Rhine to the Vistula, the league was able to control much of European trade and especially Baltic trade. By the fourteenth century, it had chapters as far north as Iceland, as far east as Riga, south to the Ukraine, and west to Venice.

For many years, the league was seen as a positive force in northern Europe. It stood up against the abuses of monarchs, stopped piracy, dredged channels, and built lighthouses. In England, league members were called Easterlings because they came from the east, and their good reputation is reflected in the word *sterling,* which comes from *Easterling* and means "of assured value."

But the league grew increasingly abusive of its power and ruthless in defense of trade monopolies. In 1381, mobs rose up in England and hunted down Hanseatics, killing anyone who could not say *bread and cheese* with an English accent.

The Hanseatics monopolized the Baltic herring trade and in the fifteenth century attempted to do the same with dried cod. By then, dried cod had become an important product in Bristol. Bristol's well-protected but difficult-to-navigate harbor had greatly expanded as a trade center because of its location between Iceland and the Mediterranean. It had become a leading port for dried cod from Iceland and wine, especially sherry, from Spain. But in 1475, the Hanseatic League cut off Bristol merchants from buying Icelandic cod.

Thomas Croft, a wealthy Bristol customs official, trying to find a new source of cod, went into partnership with John Jay, a Bristol merchant who had what was at the time a Bristol obsession: He believed that somewhere in the Atlantic was an island called Hy-Brasil. In 1480, Jay sent his first ship in search of this island, which he hoped would offer a new fishing base for cod. In 1481, Jay and Croft outfitted two more ships, the *Trinity* and the *George*. No record exists of the result of this enterprise. Croft and Jay were as silent as the Basques. They made no announcement of the discovery of Hy-Brasil, and history has written off the voyage as a failure. But they did find enough cod so that in 1490, when the Hanseatic League offered to negotiate to reopen the Iceland trade, Croft and Jay simply weren't interested anymore.

Where was their cod coming from? It arrived in Bristol dried, and drying cannot be done on a ship deck. Since their ships sailed out of the Bristol Channel and traveled far west of Ireland and there was no land for drying fish west of Ireland—Jay had still not found Hy-Brasil—it was suppposed that Croft and Jay were buying the fish somewhere. Since it was illegal for a customs official to engage in foreign trade, Croft was prosecuted. Claiming that he had gotten the cod far out in the Atlantic, he was acquitted without any secrets being revealed.

To the glee of the British press, a letter has recently been discovered. The letter had been sent to Christopher Columbus, a decade after the Croft affair in Bristol, while Columbus was taking bows for his discovery of America. The letter, from Bristol merchants, alleged that he knew perfectly well that they had been to America already. It is not known if Columbus ever replied. He didn't need to. Fishermen were keeping their secrets, while explorers were telling the world. Columbus had claimed the entire new world for Spain.

Then, in 1497, five years after Columbus first stumbled across the Caribbean while searching for a westward route to the spice-producing lands of Asia, Giovanni Caboto sailed from Bristol, not in search of the Bristol secret but in the hopes of finding the route to Asia that Columbus had missed. Caboto was a Genovese who is remembered by the English name John Cabot, because he undertook this voyage for Henry VII of England. The English, being in the North, were far from the spice route and so paid exceptionally high prices for spices. Cabot reasoned correctly that the British Crown and the Bristol merchants would be willing to finance a search for a northern spice route. In June, after only thirty-five days at sea, Cabot found land, though it wasn't Asia. It was a vast, rocky coastline that was ideal for salting and drying fish, by a sea that was teeming with cod. Cabot reported on the cod as evidence of the wealth of this new land,

New Found Land, which he claimed for England. Thirty-seven years later, Jacques Cartier arrived, was credited with "discovering" the mouth of the St. Lawrence, planted a cross on the Gaspé Peninsula, and claimed it all for France. He also noted the presence of 1,000 Basque fishing vessels. But the Basques, wanting to keep a good secret, had never claimed it for anyone.

> The codfish lays a thousand eggs
> The homely hen lays one.
> The codfish never cackles
> To tell you what she's done.
> And so we scorn the codfish
> While the humble hen we prize
> Which only goes to show you
> That it pays to advertise.
>
> —anonymous American rhyme

THE MEDIEVAL COD CRAZE

SALT COD IS EATEN WITH MUSTARD SAUCE OR WITH MELTED FRESH BUTTER OVER IT.

—Guillaume Tirel, a.k.a. Taillevent,
Le Viandier, 1375

Taillevent, master cook to Charles V of France, left this work in a rolled manuscript. Like almost every cook who came after him, he believed that salt cod was a harsh food that needed to be enriched with fat, whereas fresh cod was a bland food that needed to be enlivened with seasoning. He offered a recipe for fresh cod, as well as several for "Jance," a sauce that reflects the spice fashions of the day.

In France: Fresh Cod

Prepared and cooked like a red mullet, with wine when cooking; eaten with Jance. Some people put garlic with it, and others do not.

Jance Recipes

Cow's milk Jance: Grind ginger and egg yolks, infuse them in cow's milk, and boil.

Garlic Jance: Grind pepper, garlic and almonds, infuse them in good verjuice, then boil it. Put white wine in it (if you wish).

Ginger Jance: Grind ginger and almonds, but no garlic. Infuse this in verjuice, then boil it. Some people put white wine in it. [Verjuice was originally made from the acidic juice of sorrel and later the juice of unripened plums.]

In England: Cokkes of Kellyng (Cockles of Codling)

In this recipe, written in Middle English, a codling is cut into cockle-size pieces.

Take cokkes of kellyng; cut hem smalle. Do hit yn a brothe of fresch fysch or of fresh salmon; boyle hem well. Put to mylke and draw a lyour of bredde to hem with saundres, safferyn & sugure and poudyr of pepyr. Serve hit forth, & othyr fysch amonge: turbut, pyke, saumon, chopped & hewn. Sesyn hem with venyger & salt.

—from an anonymous manuscript in
Yale University's Beinecke Library,
dated from the twelfth century to the fifteenth
[the use of sugar argues for the fifteenth]

2: With Mouth Wide Open

IT HAS BEEN CALCULATED THAT IF NO ACCIDENT PREVENTED THE HATCHING OF THE EGGS AND EACH EGG REACHED MATURITY, IT WOULD TAKE ONLY THREE YEARS TO FILL THE SEA SO THAT YOU COULD WALK ACROSS THE ATLANTIC DRYSHOD ON THE BACKS OF COD.

—Alexandre Dumas,
Le Grande Dictionnaire de cuisine, 1873

The hero, *Gadus morhua,* is not a nice guy.

It is built to survive. Fecund, impervious to disease and cold, feeding on most any food source, traveling to shallow waters and close to shore, it was the perfect commercial fish, and the Basques had found its richest grounds. Cod should have lasted forever, and for a very long time it was assumed that it would. As late as 1885, the Canadian Ministry of Agriculture said, "Unless the order of nature is overthrown, for centuries to come our fisheries will continue to be fertile."

The cod is omnivorous, which is to say it will eat anything. It swims with its mouth open and swallows whatever will fit—including young cod. Knowing this, sports fishermen in New England and Maritime Canada jig for cod, a baitless means of fishing, where a lure by its appearance and motion imitates a favorite prey of the target fish. A cod jigger is a piece of lead, sometimes fashioned to resemble a herring, but often shaped like a young cod.

Yet cod might be just as attracted to an unadorned piece of lead. English fishermen say they find Styrofoam cups thrown overboard from Channel-crossing ferries in the bellies of cod.

The cod's greed makes it easy to catch, but the fish is not much fun for sportsmen. A cod, once caught, does not fight for freedom. It simply has to be hauled up, and it is often large and heavy. New England anglers would far rather catch a bluefish than a cod. Bluefish are active hunters and furious fighters, and once hooked, a struggle ensues to reel in the line. But the bluefish angler brings home a fish with dark and oily flesh, characteristic of a midwater fighter who uses muscles for strong swimming. The cod, on the other hand, is prized for the whiteness of its flesh, the whitest of the white-fleshed fish, belonging to the order Gadiformes. The flesh is so purely white that the large flakes almost glow on the plate. Whiteness is the nature of the sluggish muscle tissue of fish that are suspended in the near-weightless environment at the bottom of the ocean. The cod will try to swim in front of an oncoming trawler net, but after about ten minutes it falls to the back of the net, exhausted. White muscles are not for strength but for quick action—the speed with which a cod, slowly cruising, will suddenly pounce on its prey.

Cod meat has virtually no fat (.3 percent) and is more than 18 percent protein, which is unusually high even for fish. And when cod is dried, the more than 80 percent of its flesh that is water having evaporated, it becomes concentrated protein—almost 80 percent protein.

There is almost no waste to a cod. The head is more flavorful than the body, especially the throat, called a tongue, and the small disks of flesh on either side, called cheeks. The air bladder, or sound, a long tube against the backbone that can fill or release gas to adjust swimming depth, is rendered to make isinglass, which is used industrially as a clarifying agent and in some glues. But sounds are also fried by codfishing peoples, or cooked in chowders or stews. The roe is eaten, fresh or smoked. Newfoundland fishermen also prize the female gonads, a two-pronged organ they call the britches, because its shape resembles a pair of pants. Britches are fried like sounds. Icelanders used to eat the milt, the sperm, in whey. The Japanese still eat cod milt. Stomachs, tripe, and livers are all eaten, and the liver oil is highly valued for its vitamins.

Icelanders stuff cod stomachs with cod liver and boil them until tender and eat them like sausages. This dish is also made in the Scottish Highlands, where its dubious popularity is not helped by the local names: Liver-

Muggie or Crappin-Muggie. Cod tripe is eaten in the Mediterranean.

The skin is either eaten or cured as leather. Icelanders used to roast it and serve it with butter to children. What is left from the cod, the remaining organs and bones, makes an excellent fertilizer, although until the twentieth century, Icelanders softened the bones in sour milk and ate them too.

The word *cod* is of unknown origin. For something that began as food for good Catholics on the days they were to abstain from sex, it is not clear why, in several languages, the words for salt cod have come to have sexual connotations. In the English-speaking West Indies, saltfish is the common name for salt cod. In slang, *saltfish* means "a woman's genitals," and while Caribbeans do love their salt cod, it is this other meaning that is responsible for the frequent appearance of the word *saltfish* in Caribbean songs such as the Mighty Sparrow's "Saltfish."

In Middle English, *cod* meant "a bag or sack," or by inference, "a scrotum," which is why the outrageous purse that sixteenth-century men wore at their crotch to give the appearance of enormous and decorative genitals was called a codpiece. Samuel Johnson's 1755 dictionary defines *cod* as "any case or husk in which seeds are lodged." Does this have anything to do with the fish? Most scholars doubt it but offer no other explanation for the origin of the word. Henry David Thoreau conjectured that the fish was named after the husk of seeds because the female held so many millions of eggs.

There are other connections between codfish and pouches. In Quebec's Gaspé Peninsula, where the French have fished cod since before Shakespeare's birth, and where people still use every part of the fish, cod skin is cured into a kind of leather from which pouches are made. The same is done in Iceland. The fish might also be named for the pouch at the back of a net where the cod are trapped. On a modern trawler, this part of the net is still called the cod end.

In Great Britain since the nineteenth century, *cod* has meant "a joke or prank." This may be because a codpiece was presumably far larger than the parts it advertised. However, the Danish word for cod, *torsk,* also has the colloquial meaning "fool."

The French word for cod, *morue,* gave the Atlantic cod the second part of its Latin name. But curiously, sometime in the nineteenth century, while cod was becoming a prank in England, *morue* in France came to mean "a prostitute." Historic dictionaries of the French language do not offer an explanation for this, other than that it probably started with the vendors in Paris's Les Halles market who were given to such anthropomorphisms, especially with fish. Pimps were mackerel, which is an oily and predatory fish. By the nineteenth century, nothing so clearly represented unbridled commercialism as the salt cod. A *morue* is something degraded by commerce. "Yes, yes, I will desalinate you, you *grande morue*!" a character declares in Émile Zola's 1877 novel, *Assommoir.* And when Louis Ferdinand Céline wrote that the stars are "*tout morue,*" it was not that they were made of salt cod but that the universe was cheapened and perverse.

In modern French, a fresh cod is called a *cabillaud,* which comes from the Dutch *kabeljauw.* The French adopted a foreign word for the fresh fish, which did not greatly interest them, but reserved a French word, *morue,* for salt cod, which they loved. *Morue* is an older word than the word *cabillaud.* In Quebec, where the French language has barely changed since the eighteenth century, the word *cabillaud* is unknown. Quebecers speak of fresh or salted *morue.*

To the Spanish, Italians, and Portuguese, fresh cod does not even exist, and there is really no word for it. It has to be called a "fresh salt cod." Salt cod is *baccalà* in Italian and *bacalhau* in Portuguese, both of which may come from the Spanish word *bacalao.* Typical of Iberia, both the Basques and Catalans claim the word comes from their own languages, and the rest of Spain disagrees. Catalans have a myth that cod was the proud king of fish and was always speaking boastfully, which was an offense to God. "*Va callar*!" (Will you be quiet!), God told the cod in Catalan. Whatever the word's origin, in Spain, *lo que corta el bacalao,* the person who cuts the salt cod, is a colloquialism for the person in charge.

Codfish include ten families with more than 200 species. Almost all live in cold salt water in the Northern Hemisphere. Cod were thought to have developed into their current forms about 120 million years ago in the Tethys Sea, a tropical sea that once ran around the earth east–

west and connected all other oceans. Eventually the Tethys merged with a northern sea, and the cod became a fish of the North Atlantic. Later, when a land bridge between Asia and North America broke, cod found their way into the northern Pacific. In gadiform fish, evolution is seen in the fins. The cusk has almost a continuous single fin around the body with a barely distinct tail. The ling has a distinct tail and a small second dorsal fin. On a hake, the forward dorsal fin becomes even more distinct. On a whiting, there are three dorsal fins, and the anal (belly) side has developed two distinct fins. On the most developed gadiforms—cod, haddock, and pollock—these three dorsal and two anal fins are large and very separate.

Despite the warm-water origins, only one tropical cod remains: the tiny bregmaceros, of no commercial value and almost unknown habits. There is also one South Atlantic species and even one freshwater cod, the burbot, whose white flesh, though not quite the quality of an Atlantic cod, is enjoyed by lake fishermen in Alaska, the Great Lakes, New England, and Scandinavia. Norwegians think the burbot has a particularly delectable liver. There are other gadiforms that are pleasant to eat but of no commercial value. Sportsmen like to jig the coastline of Long Island and New England for the small tomcod, which also has a Pacific counterpart.

But to the commercial fisherman, there have always been five kinds of gadiform: the Atlantic cod, the haddock, the pollock, the whiting, and the hake. Increasingly, a sixth gadiform must be added to the list, the

Engraving by William Lizars from *Jardine's Naturalist's Library*, 1833.

Pacific cod, *Gadus macrocephalus*, a smaller version of the Atlantic cod whose flesh is judged of only slightly lesser quality.

The Atlantic cod, however, is the largest, with the whitest meat. In the water, its five fins unfurl, giving an elegant form that is streamlined by a curving white stripe up the sides. It is also recognizable by a square rather than forked tail and a curious little appendage on the chin, which biologists think is used for feeling the ocean floor.

The smaller haddock has a similar form but is charcoal-colored on the back where the cod is spotted browns and ambers; it also has a black spot on both sides above the pectoral fin. The stripe on a haddock is black instead of white. In New England, there is a traditional explanation for this difference. There, cod is sometimes referred to as "the sacred cod." In truth, this is because it has earned New Englanders so many sacred dollars. But according to New England folklore, it was the fish that Christ multiplied to feed the masses. In the legend, Satan tried to do the same thing, but since his hands were burning hot, the fish wriggled away. The burn mark of Satan's thumb and forefinger left black stripes; hence the haddock.

This story illustrates the difference, not only in stripes but in status, between cod and haddock. British and Icelandic fishermen only reluctantly catch haddock after their cod quotas are filled, because cod always brings a better price. Yet Icelanders prefer eating haddock and rarely eat cod except dried. Asked why this is so, Reykjavík chef Úlfar Eysteinsson said, "We don't eat money."

The stars are *tout morue*, and cod is money; haddock is simply food. The Nova Scotians, true to their namesakes, prefer haddock, even for fish-and-chips, which would be considered a travesty in Newfoundland and virtually a fraud in the south of England. In the north of England, as in Scotland, haddock is preferred.

In places far from the range of Atlantic cod, hake is a substitute. The rare gadiform that is found in both the Northern and Southern Hemispheres, hake is a popular fish, fresh and cured, off of Chile, Argentina, New

Zealand, and especially South Africa. Basques, who prize salt cod above all other fish, would rather eat a fresh hake than a fresh cod, which few have ever even seen. Because hake is found in waters closer to Spain, including the Mediterranean, *cod* has come to mean "cured," while *hake* means "fresh." Some Basque chefs say they prefer hake tongues to cod tongues, but what they are really saying is they prefer fresh tongues to cured ones.

Cod is the fish of choice for curing, though all of the other gadiforms are cured too, often now as a less costly substitute for cod. Salt ling is a Scottish tradition, and speldings, wind-dried whiting wetted with seawater as they dried to give a special taste, became a local specialty north of Aberdeen in the eighteenth century. At the same time, south of Aberdeen, haddocks were being dried on shore and smoked over peat and seaweed fires by the wives of the fishermen of Findon—which is the origin of the still-celebrated finnan haddie. This has achieved such status that an occasional bogus smoked cod is passed off in the United States as finnan haddie, while a salted haddock might be passed off as salt cod.

But in spite of the occasional local preference, on the world market, cod is the prize. This was true in past centuries when it was in demand as an inexpensive, long-lasting source of nutrition, and it is true today as an increasingly expensive delicacy. Even with the Grand Banks closed, worldwide more than six million tons of gadiform fish are caught in a year, and more than half are *Gadus morhua,* the Atlantic cod. For fishermen, who are extremely tradition bound, there is status in fishing cod.

Proud cod fishermen are indignant, or at least saddened, by the suggestion that they should switch to what they see as lesser species.

In addition to its culinary qualities, the cod is eminently catchable. It prefers shallow water, only rarely venturing to 1,800 feet, and it is commonly found in 120 feet (twenty fathoms) or less. Cod migrate for spawning, moving into still-shallower water close to coastlines, seeking warmer spawning grounds and making it even easier to catch them.

They break off into subgroups, which adapt to specific areas, varying in size and color, from yellow to brown to green to gray, depending on local conditions. In the dark waters off of Iceland, they are brown with yellow specks, but it takes only two days in the brightly lit tank of an aquarium in the Westman Islands, off of Iceland, for a cod to turn so pale it looks almost albino. The so-called northern stock, the cod off of Newfoundland and Labrador, are smaller for their age than the cod off of Massachusetts, where the water is warmer. Though always a cold-water fish, preferring water temperatures between thirty-four and fifty degrees, cod grows faster in the warmer waters of its range. Historically, but not in recent years because of overfishing, the cod stock off of Massachusetts was the largest and meatiest in the world.

Cod manufacture a protein that functions like antifreeze and enables the fish to survive freezing temperatures. If hauled up by a fisherman from freezing water, which rarely happens since they are then underneath ice, the protein will stop functioning and the fish will instantly crystallize.

Cod feed on the sea life that clusters where warm and cold currents brush each other—where the Gulf Stream passes by the Labrador current off North America, and again where it meets arctic currents off the British Isles, Scandinavia, and Russia. The Pacific cod is found off of Alaska, where the warm Japanese current touches the arctic current. In fact, the cod follow this edge of warm and cold currents so consistently that some scientists believe the shifting of weather patterns can be monitored by noting where fishermen find cod. When cold northern waters become too cold, the cod populations move south, and in warmer years they move north.

From Newfoundland to southern New England, there is a series of shallow areas called banks, the southernmost being Georges Bank off of Massachusetts, which is larger than the state. Several large banks off of Newfoundland and Labrador are together called the Grand Banks. The largest of the Grand Banks, known as the Grand Bank, is larger than Newfoundland. These are huge shoals on the edge of the North American continental shelf. The area is rich in phytoplankton, a growth produced from the nitrates stirred up by the conflicting currents. Zooplankton, tiny sea creatures, gorge themselves on the phytoplankton. Tiny shrimplike free-floating creatures called krill eat the zooplankton. Herring and other midwater species rise to eat the krill near the surface, and seabirds dive for both the krill and the fish. Humpback whales also feed on krill. And it is this rich environment on the banks that produces cod by the millions. In the North Sea, the cod grounds are also found on banks, but the North American banks, where the waters of the Gulf of Mexico meet the arctic Greenland waters, had a greater density of cod than anything ever seen in Europe. This was the Basques' secret.

Still more good news for the fishermen, a female cod forty inches (102 centimeters) long can produce three million eggs in a spawning. A fish ten inches longer can produce nine million eggs. A cod may live to be twenty or even thirty years old, but it is the size more than the age that determines its fecundity. Dumas's image of all the eggs hatching so that someone could walk across the ocean on the backs of cod is typical nineteenth-century enthusiasm about the abundance of the species. But it could never happen. In the order of nature, a cod produces such a quantity of eggs precisely because so few will reach maturity. The free-floating eggs are mostly destroyed as they are tossed around the ocean's surface, or they are eaten by other species. After a couple of weeks, the few surviving eggs hatch and hungrily feed, first on phytoplankton and soon zooplankton and then krill. That is, if they can get to those foods before the other fish, birds, and whales. The few cod larvae that are not eaten or starved in the first three weeks will grow to about an inch and a half. The little transparent fish, called juveniles, then leave the upper ocean and begin their life on the bottom, where they look for gravel and other rough surfaces in which to hide from their many predators, including hungry adult cod. A huge crop of eggs is necessary for a healthy class, as biologists call them, of juveniles. If each female cod in a lifetime of millions of eggs produces two juveniles that live to be sexually mature adults, the population is stable. The first year is the hardest to survive. After that, the cod has few predators and many prey. Because a cod will eat most anything, it adapts its diet to local conditions, eating mollusks in the Gulf of Maine, and herring, capelin, and squid in the Gulf of St. Lawrence. The Atlantic cod is particularly resistant to parasites and diseases, far more so than haddock and whiting.

If ever there was a fish made to endure, it is the Atlantic cod—the common fish. But it has among its predators man, an openmouthed species greedier than cod.

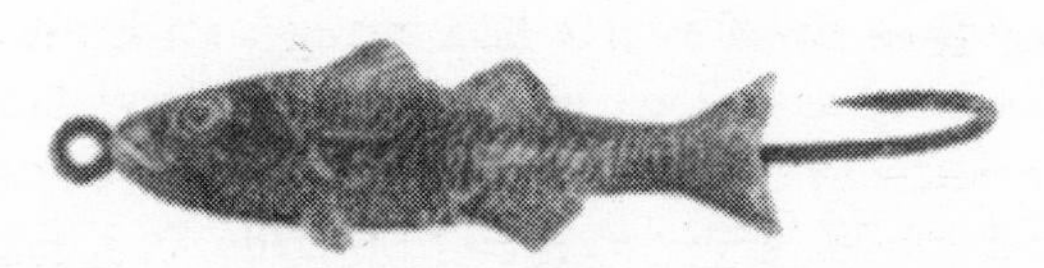

THE WELL-COOKED HEAD

Hannah Glasse's recipes show how much has been lost from the craft of British cooking, especially the art of roasting. A century after Glasse, French food writer Jean Anthelme Brillat-Savarin wrote, "You may be born to cook, but you must learn to roast."

To Roast a Cod's Head

Wash it very clean, and Score it with a Knife, strew a little Salt on it, and lay it in a Stew-pan before the Fire, with something behind it, that the Fire may Roast it. All the Water that comes from it the first half Hour, throw away; then throw on it a little Nutmeg, Cloves, and Mace beat fine, and Salt; flour it, and baste it with Butter. When that has lain Some time, turn it, and season, and baste the other side the same; turn it often, then baste it with butter and Crumbs of Bread. If it is a large Head, it will take four or five Hours baking; have ready some melted Butter with an Anchovy, some of the Liver of the Fish boiled and bruised fine, mix it well with the Butter, and two yolks of Eggs beat fine, and mixed with the Butter, then strain them through a Sieve, and put them into the sauce pan again, with a few Shrimps, or pickled Cockles, two Spoonfuls of Red Wine, and the Juice of a Lemon. Pour it into the Pan the head was roasted in, and stir it all together, pour it into the Sauce-

pan, keep it stirring, and let it boil; pour into a Bason. Garnish the Head with fried Fish, Lemon, and scraped Horse-reddish. If you have a large Tin Oven it will do better.

—Hannah Glasse,
The Art of Cookery: Made PLAIN and EASY which far exceeds any Thing of the Kind ever yet Published BY A LADY, London, 1747

Glasse also offered equally elaborate recipes for both boiled and baked cod head.

Also see pages 241–44.

3: The Cod Rush

IF CODFISH FORSAKE US, WHAT THEN WOULD WE HOLD?
WHAT CARRY TO BERGEN TO BARTER FOR GOLD?

—Peter Daas,
Trumpet of Nordland, Norway, 1735

The Basque secret was out.

Raimondo di Soncino, Milan's envoy in London, had written a letter to the duke of Milan on December 18, 1497, reporting John Cabot's return on August 6:

> The Sea there is swarming with fish which can be taken not only with the net but in baskets let down with a stone, so that it sinks in the water. I have heard

> this Messer Zoane state so much. These same English, his companions, say that they could bring so many fish that this Kingdom would have no further need of Iceland, from which there comes a very great quantity of the fish called stockfish.

From this statement, historians have concluded that John Cabot's men caught cod simply by dropping weighted baskets. There is no evidence that Cabot ever said this, nor is it known how reliable di Soncino's source was. However, subsequent accounts do confirm that the coast of North America was churning with codfish of a size never before seen and in schools of unprecedented density, at least in recorded European history.

When Europeans first arrived, North America had a wealth of game and fish unparalleled in Europe. Flocks of birds, notably the passenger pigeon, which is now extinct, would darken the sky for hours as they passed overhead. In 1649, Adriaen van der Donck, the colonial governor of New Amsterdam, wrote from what is now New York that nearby waters had six-foot lobsters. Even a century after Cabot, Englishmen wrote of catching five-foot codfish off Maine, and there are persistent accounts in Canada of "codfish as big as a man." In 1838, a 180-pounder was caught on Georges Bank, and in May 1895, a six-foot cod weighing 211 pounds was hauled in on a line off the Massachusetts coast. Cabot's men may well have been able to scoop cod out of the sea in baskets.

Cabot, a skilled and experienced navigator, had moved to Bristol with his wife and son only two years before his 1497 voyage, frustrated by the triumphs of Columbus and dreaming of his own glory. Both Columbus and Cabot had been born in Genoa about the same year, and both had searched the Mediterranean for backers. They probably knew each other. Cabot may have even had to endure the spectacle of some of Columbus's triumphant receptions. He seems to have been in Barcelona in April 1493, when crowds cheered his fellow Genovese formally entering the city. At last, when Cabot returned to England after his North American voyage, he basked in the same kind of reception that Columbus had enjoyed in Spain. In England, Cabot was a sensation, the man of the moment, and fans assailed him on the streets of Bristol and London the way they might today if he were a rock star. But there was little time to luxuriate in what might be only fleeting glory. After all, Columbus was about to embark on his *third* voyage. With his sudden fame, Cabot easily raised funding for a second voyage with five ships. One ship soon returned, and the other four, along with Cabot, were never heard from again. It was the first of many such calamities.

The Portuguese were also exploring and charting North America. A 1502 map identifies Newfoundland as "land of the King of Portugal," and to this day, many Portuguese consider Newfoundland to be a Portuguese "discovery." Many of the earliest maps of Newfoundland show Portuguese names. Those names have remained, though they are no longer recognizably Portuguese. Cabo de Espera (Cape Hope), the tip of land between St. John's and Petty Harbour, has become Cape Spear, Cabo Raso is now Cape Race, and the Isla dos Baccalhao is Baccalieu Island. In 1500, Gaspar Côrte Real went to Newfoundland and named it Greenland, Terra Verde. He was the youngest son of João Vaz Côrte Real, a despotic ruler of the Azores and yet another mariner who some claim went to America before Columbus. In 1501, on his second trip, after sending back fifty-seven Beothuk as sample slaves, Gaspar, like Cabot, vanished with his ship and crew. The following year, his brother Miguel was lost along with his flagship and its crew.

This grim early record did not discourage fishermen. Fishing had opened up in Newfoundland with the enthusiasm of a gold rush. By 1508, 10 percent of the fish sold in the Portuguese ports of Douro and Minho was Newfoundland salt cod. In France, the Bretons and Normans had an advantage because the profitable markets of the day were nearby Rouen and Paris. By 1510, salt cod was a staple in Normandy's busy Rouen market. By midcentury, 60 percent of all fish eaten in Europe was cod, and this percentage would remain stable for the next two centuries.

The sixteenth-century Newfoundland cod trade was changing markets and building ports. La Rochelle on the French Atlantic coast had been a second-string harbor because it was not on a river, a critical flaw since goods were moved on rivers. All La Rochelle had, in addition to a well-protected harbor, was a determined Protestant merchant class that saw the commercial opportunity in Newfoundland cod. Yet La Rochelle became the premier

Newfoundland fishing port of Europe. Of the 128 fishing expeditions to Newfoundland between Cabot's first voyage and 1550, more than half were from La Rochelle.

The French dominated these years, originating 93 of those 128 fishing expeditions to Newfoundland. The rest were divided between the English, Spanish, and Portuguese. Figures on the Basques, as is the Basque fate, are buried in French and Spanish statistics, but the French Basque ports of Bayonne and St.-Jean-de-Luz were important in the first half of the century.

Even though Cabot had claimed North America for England, British fishermen had not immediately joined the cod rush because catches were good in Iceland. It was cod that had first lured Englishmen from the safety of their coastline in pre-Roman times. By the early fifteenth century, two- and three-masted ketches with rudders were going to Iceland and the Faroes. Not only were these some of the best fishing vessels of the day, but not until the twentieth century would Icelanders have vessels of an equal quality for fishing their own waters.

But the conflict between England and the Germans of the Hanseatic League over rights to Icelandic cod grew steadily worse. In 1532, an Englishman, John the Broad, was murdered in the Icelandic fishing station of Grindavík. Though Britain's Icelandic Cod Wars are thought of as a twentieth-century phenomenon, the first one was set off by this Grindavík killing and was fought not against Iceland, which was a colonized and docile nation by then, but against the Hanseatic League, which had developed its own navy. Uncharacteristic of the British,

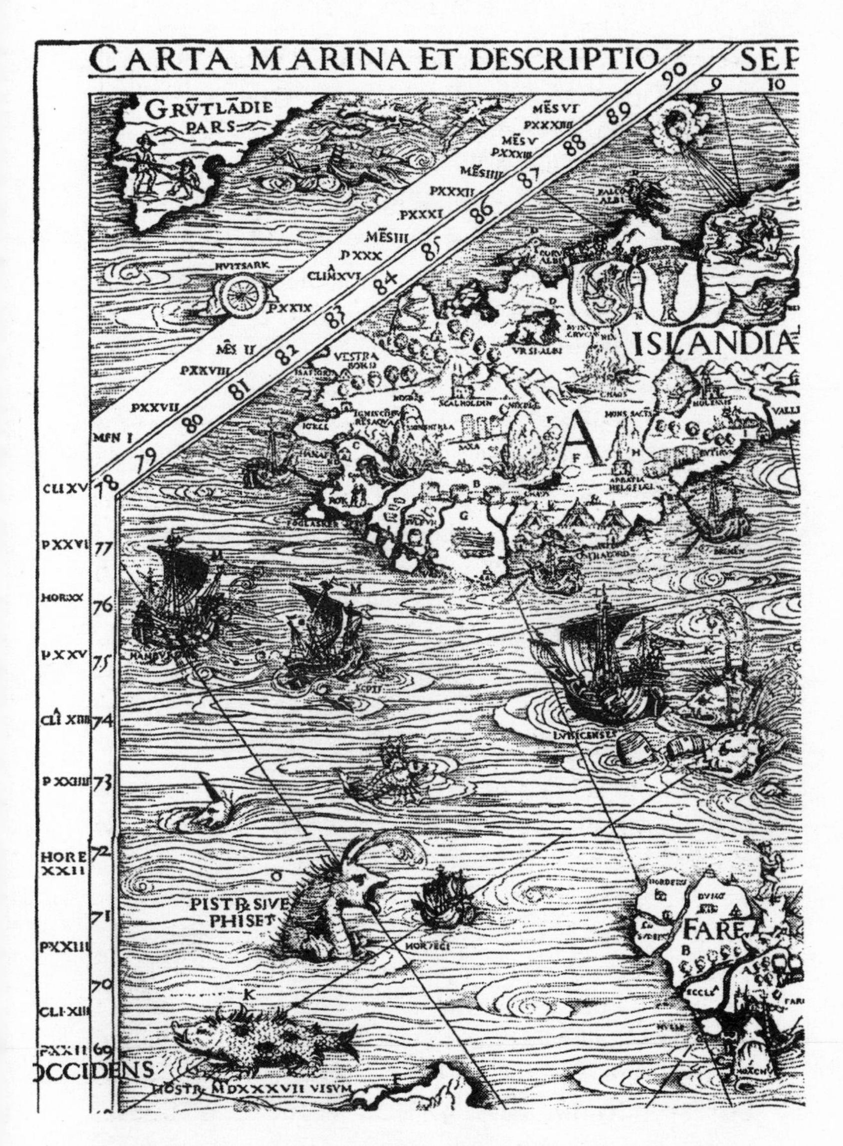

Detail showing Cod War of 1532 off of Grindavík from Olaus Magnus's *Carta Marina*, 1539. (Uppsala University Library, Uppsala)

after a brief fight they simply withdrew from the Icelandic fishery. As di Soncino had predicted, Britain didn't need Iceland anymore.

With the opening up of Newfoundland, the British West country began developing major fishing ports. In the days of slow sailing, a westward location was a tremendous advantage because it reduced the length of a voyage. Except Ireland, which was too impoverished to develop a distant water fleet, the ports that remained important to the Newfoundland fishery into the mid-twentieth century—St.-Malo on the Brittany peninsula, Vigo on the northwest tip of Spain, the Portuguese ports—were those in the European regions closest to Newfoundland.

The Spanish Basque city of Bilbao, with its ironworks providing the anchors and other metal fittings for Europe's ships, was one of the ports that grew with the boom in shipbuilding created by the cod trade. According to historian Samuel Eliot Morison, at no time in history, not even during World War II, has there ever been such a demand for replacement of sunken ships as between 1530 and 1600. European ambition was simply too far ahead of technology, and until better ships and better navigation were developed, shipwrecks and disappearances were a regular part of this new adventure.

In this rapidly expanding commercial world, the British had one great disadvantage over the French, Spanish, and Portuguese: They had only a modest supply of salt. Most northern countries lacked salt and simply produced winter fish that was dried without salting. It was called stockfish, from the Dutch word *stok*, meaning "pole," because the fish were tied in pairs by the tail and hung over poles to dry, as is still done out on the lava fields of Iceland every winter.

But the English wanted to produce a year-round supply of cod for a growing market, and since neither the North Sea nor Iceland was cold enough for drying fish in the summer, they became dependent on salting. Some fish were simply sold salted and undried, which became known as "green" not because of the color but because it was considered a more natural state than dried fish. But in an attempt to conserve their limited salt, the British invented a product that was to be favored in Mediterranean and Caribbean markets for centuries: a lightly salted dried cod. The Norwegians called it *terranova fisk*, Newfoundland fish, but later used the name *klipfisk*, rockfish, because it was dried on rocky coasts.

As green and salted-and-dried fish became available, they were preferred to the unsalted stockfish and brought substantially higher prices. The British experimented with new products such as a summer-cured dried cod from the Grand Banks known as Habardine or Poor John. In Shakespeare's *The Tempest*, Trinculo says of Caliban, whom he finds on the beach, "He smells like a fish—a very ancient and fish-like smell—a kind of, not of the newest, poor john."

Winter cures were known to be superior. Other variations were developed. Some fish was salted directly, and some was pickled in brine in barrels. Some of the pickled and some of the green were later dried to give them more durability. There was not only a wide choice of products in cured Newfoundland cod but also, no doubt, a great range of quality. "As to their Quality, Many of them Stink, for 'tis a certain Maxim, that if Fish or Flesh be not well cured and salted first, they cannot be recovered," John Collins, an accountant to the Royal Fishery, wrote in *Salt and Fisheries.*

It is not by chance that a Royal Fishery accountant was publishing a book on salt in 1682. The British fisheries had by then been wrestling with the salt problem for centuries. Collins pointed out that brackish water around England could be boiled, which yielded more salt than did evaporating seawater. He discussed the relative quality of salt and offered this recipe for one of the better English salts.

> . . . the manner of boyling the Brine into Salt at Namptwich. They boyl it in Iron Pans, about 3 foot square, and 6 inches deep; their Fires are made of Staffordshire Pit-Coles, and one of their smaller Pans is boiled in 2 hours time.
>
> To clarify and raise the Scum, they use Calves, Cows and Sheeps blood, which in *Philosophical Transaction, No 142,* is said to give the Salt an ill flavour.

Wich is an Anglo-Saxon word meaning "a place that has salt," and all the English towns whose names end in *wich* were at one time salt producers. But they could never produce enough for the Newfoundland cod fishery.

Collins warned against French salt, which he said was

unhealthy. He probably had several reasons for saying this, aside from a general dislike of the French. There was a great tradition of French contraband, and salt was a favorite item because the French had a near obsession with evading the salt tax and, in fact, most taxes. "Oh, the rain washed it away" or "Someone must have stolen it" was the familiar litany recited to salt tax collectors. The British also had a hated salt tax and had their homes searched for off-the-books salt. But the French salt tax, the *gabelle,* was particularly hated and was one of the grievances leading to the French Revolution. Like many reforms of the Revolution, the abolition of the salt tax lasted only fifteen years and then was reinstated until 1945. One way of getting around the salt tax was to make your own, by boiling brackish water, and probably much of this illicit salt smuggled to England was indeed, as Collins said, unhealthy.

But the French Terreneuve merchants filled their holds with legal, high-quality French salt, which made good ballast, and sailed to Newfoundland. They returned with salt cod in the holds where the salt had been.

Salt was a great advantage of the Bretons. Under the agreement by which the duchy of Brittany became part of France, Bretons were exempt from the *gabelle.* And since sixteenth-century salt was made from evaporation, it was a southern product and southern Brittany was the most northerly point in western Europe where salt making was commercially viable.

Nearby Brittany could have supplied the British salt needs, but the French were the enemy. It was Portugal, with its saltworks in Aveiro—which, not by coincidence, became and still is Portugal's salt cod center—that had what was considered Europe's best salt. Bristol merchants went into a number of joint ventures with the Portuguese. In exchange for salt, the British government gave Portuguese ships protection from the French. In 1510, the king of Portugal complained to the king of France that French ships had taken 300 Portuguese vessels in the past ten years.

The mutually advantageous British arrangement with the Portuguese lasted until 1581, when Portugal merged with Spain. It was a bad moment for a seagoing nation to throw in its lot with Spain. In 1585, the British attacked and destroyed the Spanish fishing fleet, and the military fleet was destroyed in its disastrous attempt to invade England. The Spanish fleets took the Portuguese down with them. The Portuguese continued to fish the Grand Banks until expelled by the Canadian government in 1986, but after their short-lived merger with Spain ended their British alliance, they were never again a dominant force in the Newfoundland fishery.

By the time England broke its alliance with Portugal, not quite a century after Cabot's first voyage, Newfoundland cod was more than commerce to the British; it was strategic. In fact, what finally spurred the British to become the dominant players in the Newfoundland fishery in the second half of the sixteenth century was providing enormous quantities of dried—not salted—cod to the British Navy's ships-of-the-line fighting France. They fed their Navy with it and sold the surplus. Quick to catch fish, the English were slow to learn the European market and had trouble selling their fish to Mediterranean countries where the population demanded high-quality salted and dried fish. After a century-long free-for-all, the Spanish Basques were reemerging as dominant suppliers to the Mediterranean world—despite losing their secret.

British law greatly encumbered its own attempts at trade. Since Newfoundland cod was strategic, its commerce had to be tightly controlled, as though cod were a weapon of war. The Spanish and Portuguese had also viewed cod as strategic, because it sustained the crews on their increasing number of tropical voyages to the New World. But the Iberians also had enormous home markets for cured cod. England had the smallest market for cured cod of any of the cod-fishing nations. The English, who ate less fish, had their own highly developed home fisheries. Yet the British Crown inhibited foreign trade in cod, forbidding British ships to sell directly to European ports.

The British were landing what for that time was enormous quantities of fish. Western ports continued to grow. Plymouth on the Cornish peninsula, stretching west toward the new lands, became increasingly important. There were fifty Newfoundland fishing ships based in Plymouth alone, about which Sir Walter Raleigh wrote in 1595, "If these should be lost it would be the greatest blow ever given to England."

In 1597, this fifty-ship fleet returned from the Grand Banks, sailing up the south Cornish coast to Plymouth. It was a sight unknown in our age—some 200 canvas sails crowding the sky as the fleet made its way into the sheltered harbor against the green patchwork hills of Devon. These two-masted ships on which dozens of men lived and worked for months were only 100 feet long. Merchants would have preferred bigger ships with bigger holds, but sailors wanted to navigate the treacherous new rock-bound world with small vessels. Merchants from Holland, France, and Ireland packed into the small port town, waiting for the Plymouth fleet so they could buy their fish and ship it out again to Europe's markets.

The British continued to miss out on this commercial opportunity. In 1598, a Newfoundland fleet sailed into Southampton and sold most of the cod to French merchants, who resold them to Spain. By then, Catholic religious wars with the Huguenots, the Protestants in La Rochelle, reduced the French fleet. With Portuguese, Spanish, and French fleets all in decline, the British began to understand the commercial potential of their Newfoundland fishery. By the end of the sixteenth century, British ships were finally allowed to take their Newfoundland cod directly to foreign ports. The newly freed British traders forced open commerce in cod, and other trade followed.

But this opening of trade would seem minor in hindsight, because early in the seventeenth century, when it was just beginning, an even more important change in world trade was seeded. A small group of religious dissidents who had fled England were staring at a map in their Dutch refuge and had noticed a small hook of land that was labeled with an intriguing name—Cape Cod.

THE SHAME OF IT

Stockfish

Beat it soundly with a Mallet for half an hour or more and lay it three days a soaking, then Boyle it on a simmering Fire about an hour, with as much water as will cover it till it be soft, then take it up, and put in butter, eggs, and Mustard champed together, otherwise take 6 potato (which may be had all the year at Seed-Shops;) boyl them very tender, and then skin them. Chop them, and beat up the Butter thick with them, and put it on the fish and serve them up. Some use Parsnips.

The like for Haberdine and Poor-Jack, I should be ashamed of this Receipt if we had no better to follow, and think it too mean to mention any thing about Green Fish or barreld Cod, but the watering and soaking before they are boyled.

—John Collins,
Salt and Fishery, London, 1682

Also see pages 237–41.

4: 1620: The Rock and the Cod

FISHIEST OF ALL FISHY PLACES WAS THE TRY POTS, WHICH WELL DESERVES ITS NAME; FOR THE POTS THERE WERE ALWAYS BOILING CHOWDERS. CHOWDER FOR BREAKFAST, AND CHOWDER FOR DINNER, AND CHOWDER FOR SUPPER, TILL YOU BEGIN TO LOOK FOR FISH-BONES COMING THROUGH YOUR CLOTHES. THE AREA BEFORE THE HOUSE WAS PAVED WITH CLAMSHELLS. MRS. HUSSEY WORE A POLISHED NECKLACE OF CODFISH VERTEBRA; AND HOSEA HUSSEY HAD HIS ACCOUNT BOOKS BOUND IN SUPERIOR OLD SHARK-SKIN. THERE WAS A FISHY FLAVOR TO THE MILK, TOO, WHICH I COULD NOT ACCOUNT FOR, TILL ONE MORNING HAPPENING TO TAKE A STROLL ALONG THE BEACH AMONG SOME FISHERMEN'S BOATS, I SAW HOSEA'S BRINDLED COW FEEDING ON FISH REMNANTS, AND MARCHING ALONG THE SAND WITH EACH FOOT IN A COD'S DECAPITATED HEAD, LOOKING VERY SLIPSHOD, I ASSURE YE.

—Herman Melville, on Nantucket,
from *Moby Dick or The Whale,* 1851

For Europeans, the known world doubled in the course of the sixteenth century. The Dutch had two possibilities to offer the English Puritan refugees: the tiny, well-protected port on the tip of the island of Manhattan, or Guiana, on the shoulder of South

America. Of the two, Guiana seemed to offer better opportunities.

More than fifty years earlier, a Spaniard named Juan Martinez had been sent there as a death sentence. He had been found responsible for an accident in which a magazine of gunpowder exploded, and his punishment for negligence was to be dropped off at the unknown northeastern coast of South America in a canoe without supplies. His canoe drifted into the hands of local tribesmen, who blindfolded this first European they had ever seen and brought him to a magnificent city of palaces. After seven months, they loaded Martinez with gold and sent him on his way, again blindfolded. That, at least, was Martinez's story when he arrived in Trinidad.

Martinez's fabled city became popularly known as El Dorado. He died soon after in Puerto Rico, and the cause of death, according to many throughout the centuries, was that the gold of Guiana is cursed. In the early 1600s, it was already known that hunts for this El Dorado, even by the great Sir Walter Raleigh, had always ended disastrously. But then, so many voyages to the New World did.

With the world so greatly expanded and seemingly so empty and unknown, searching had become a European passion. Provisioned with nourishing cured cod, some headed to South America looking for gold. Others went to North America looking for cod. But what most who went to either place were really still looking for was Asia. In the sixteenth century, Newfoundland was charted as an island off of China. Europeans had sailed as far south as Maine's Bay of Fundy and not found a passage. The Spanish and Portuguese had worked south from Florida to the subantarctic tip of Patagonia and had not gotten through either. Still, the idea wouldn't die.

With the backing of Lyons silk merchants, the French Crown commissioned a Florentine, Giovanni da Verrazzano, to search for a short westward route to China. But France's Italian failed, just as Spain's and England's had. In his 1524 voyage, Verrazzano turned north, following an endless coastline from Cape Fear in present-day North Carolina. He noted that the indigenous people were "nimble and great runners," optimistically pointing out that he understood this to be characteristic of people in China. He sailed up the coast, found New York Harbor, Narragansett Bay, and an arm-shaped hook of land, which he named Pallavisino after an Italian general. Then he continued up to the coast of Maine, which he called Land of Bad People, and on to what he called "the land that in times past was discovered by the British." Having exhausted his supplies and found "7000 leagues of new coastline" and still no passage, he gave up and returned to France, where he insisted that there was a whole New World out there.

But ideas are not easily conquered by facts, and seventy-eight years later Bartholomew Gosnold was still looking for the passage to Asia. In 1602, Gosnold sailed beyond Nova Scotia, following the coast south to New England in search of a passage to Asia, where he intended to gather sassafras, which was highly prized because it was thought to cure syphilis.

North America abounds in sassafras. The Native Americans used the leaves to thicken soups. It never cured syphilis, but the roots made an excellent drink later known as root beer. Gosnold did not find China, but he returned to England with sassafras. "The powder of the sassafras," reported one of his officers, "cured one of our company that had taken a great surfeit by eating the bellies of dogfish—a very delicious meat."

Europeans still did not understand how large this thing was, this obstacle in the way of Asia. Gosnold's seemingly unsuccessful 1602 voyage ended up in history books chiefly because he renamed Pallavisino, as Cape Cod, after having reported that his ship, while in pursuit of Asian sassafras, was constantly being "pestered" by these fish.

> In the moneths of March, April and May, there is upon this coast better fishing, and in as great plentie as in Newfoundland. . . . And besides the places were but in seven faddomes water and within less than a league of the shore, where in Newfoundland they fish in fortie or fiftie fadome water and farre off.

From the European point of view in that "age of discovery," Gosnold had "discovered" New England. Yes, it had been discovered before, but in more than seventy-five years no one had been interested in Pallavisino. Gosnold's name for it, Cape Cod, like the name El Dorado in the South, opened up this new territory.

In 1603, Bristol merchants checked out Gosnold's

story and reported not only plentiful cod but excellent rocky coastline for drying fish in what is now Maine. One merchant, George Waymouth, after seeing the Maine coast, reported "huge, plentiful cods—some they measured to be five foot long and three foot about." The fact that he also confirmed the presence of sassafras seemed to get lost.

The new area was called North Virginia. In 1607, an attempt to establish a settlement there, near what is today Brunswick, Maine, resulted in the first New England–built seagoing vessel, constructed by the colonists in order to flee for England after enduring one winter. North Virginia was "over-cold," they explained, and uninhabitable.

Gosnold's map vanished, but John Smith either had seen it or at least knew some details of his voyage. By the time the Pilgrims were making their decision, Captain John Smith was already a well-known figure, in part for establishing a colony in the lower part of Virginia, but even more for his 1614 voyage to over-cold North Virginia, where he became rich from cod. Smith had actually hoped to get rich from whales, gold, and copper. He no more found any of these than Raleigh had found gold or Gosnold China. So Smith busied his crew filling the ship's hold with salt cod. He openly disliked fishing and left his men to it while he went off with a small crew in a little open boat to explore the coast. He had previously done this over 3,000 miles of inlets in the Chesapeake Bay. He now charted the coastline from Penobscot Bay in Maine to Cape Cod, making a map that included twenty-five "excellent good harbors" that he had sounded. For some reason, Gloucester's harbor was not among them.

Smith did chart the cape on which Gloucester now sits and named it after a Turkish woman of whom he had fond memories from his days soldiering in Turkey. But when he returned to England with the map, Prince Charles renamed the cape after his own mother and it has been Cape Ann ever since. Smith named several spots after Turkish memories, though none of those names remains. His new name for North Virginia, New England, proved more enduring.

Smith returned with 7,000 green cod, which he sold in England, and 40,000 stockfish, which, now that England had opened up trade with Europe, he sold in Malaga. According to Massachusetts governor William Bradford's chronicles, the Pilgrims heard Smith had done even better—that he sold 60,000 cod. Not mentioned by any of the Puritans were Smith's additional profits from twenty-seven native locals, whom he lured onto his ship and trapped in the hold until they could be sold as slaves in Spain.

In 1616, Smith published his map and a description of New England in the hope of interesting prospective settlers. And so, studying the famous captain's map, the Pilgrims decided to ask England for a land grant to North Virginia, where there was this Cape Cod. Bradford wrote, "The major part inclined to go to Plymouth, chiefly for the hope of present profit to be made by the fish that was found in that country." When the British court asked them what profitable activity they could engage in with a land grant, they said fishing.

Of all the unlikely American success stories of the epoch, none is more improbable than that of the Pilgrims. They set sail to pursue their religion and live on fishing in a new world. The fact that they arrived at the onset of winter is the first hint of how little they knew about survival. Still, they had gone to New England for fishing and not farming, and though doubtless they had never thought about this, New England does have the good fortune, unlike Newfoundland, to have a winter inshore fishing season. So why were the Pilgrims starving in the richest fishing grounds ever recorded?

It seems these religious zealots had not thought to bring much fishing tackle—not that they would have known how to use it. They knew nothing about fishing. Four or five English ships fished in New England in 1616. The year after the Plymouth landing, 1621, while the Pilgrims were nearly starving, ten British ships were profitably fishing cod in New England waters. The following year, thirty-seven ships were sent. By 1624, fifty British fishing ships were working off the coast.

Part of the Pilgrims' problem was that settlers kept coming: Thirty-five arrived the second year, and another sixty-seven in 1622. These were people unified by their religious zeal. Not only couldn't they fish, they didn't know how to hunt. They were also bad at farming. In fact, they never had a good harvest until they learned to fish cod and plow the waste in the ground as fertilizer. Their greatest food-gathering skill in the early years seems to have been an ability to find huge food caches hidden away by the native tribesmen.

What made it worse, being English, they did not want to eat unfamiliar food. The native New Englanders were Naumkeag of the Massachusetts nation. *Naumkeag,* a word meaning "fishing place," was their name for the region. The Naumkeag made line and nets from vegetable fibers and hooks from bones. In addition to catching cod and other fish that approached the coastline, they harpooned six-foot-long sturgeons in the rivers, caught eels, and delighted in the clams they harvested from the shore. They showed the Pilgrims how to pry open the big hard-shelled quahogs and the smaller thin-shelled ones, their favorites, which today New Englanders call steamers.

"Oh dear," said the Pilgrims with horrified faces. They would not eat such things. The abundant mussels, too, were rejected and continued to be shunned by New Englanders until the 1980s. The waters were so rich in lobsters that they were literally crawling out of the sea and piling up inhospitably on the beaches. But the Pilgrims, and most people until this century, did not want to eat these huge, clacking, speckled sea monsters. Apparently in desperation, they were eventually reduced to eating lobster. In 1622, Bradford reported with shame that conditions were so bad for the settlers that the only "dish they could presente their friends with was a lobster."

So what exactly did these people, not known for their open-mindedness, want to eat? The Naumkeag nickname for them was *kinshon,* which means "fish." The Pilgrims did not have skills, but they had determination. In 1623, they established a fishing station in Gloucester, which failed. They tried again two years later with little more success. But they sent back to England for equipment and advice, and with the help of Englishmen whom they referred to as "merchant adventurers," they gradually became fishermen. Fishing stations were established in Salem, Dorchester, Marblehead, and Penobscot Bay. Tidal pools were filled with seawater to make salt for the fishery. They traded chiefly with the Spanish Basque port of Bilbao, and soon they were returning with Spanish salt in their holds. They also started trading with the sugar colonies of the West Indies, and as early as 1638 a ship brought salt from the Tortugas. Through trade, New Englanders were solving the salt problem that the British never had been able to resolve through diplomacy.

In Salem, a Puritan minister, Francis Higgenson, wrote in 1629, "The aboundance of sea fish are almost beyond believing." By the end of the century, Salem was to gain more enduring fame for its mass hysteria, but originally it was noted for cod fishing. When the infamous Court of Oyer and Terminer was established in 1692, interrogating hundreds of women for witchcraft, and hanging nineteen of them, a codfish was on its seal.

In 1640, barely a generation after dreaming that Smith's 47,000 fish were a fantastic 60,000 fish, the Massachusetts Bay Colony brought 300,000 cod to the world market.

The distinct destinies that would make the northern

Cod drying on fish flakes, Marblehead, Massachusetts, detail from "View of Skinner's Head" from *Gleason's Pictorial,* vol. 6, 1854. (Peabody Essex Museum, Salem, Massachusetts)

lands Canada's eastern provinces and the southern ones the United States' New England states were established. The fundamental difference was climate. Newfoundland, the Grand Banks, the Gulf of St. Lawrence were all fished in the summertime. The ships would leave Europe in April, when there was a good easterly wind. Trans-Atlantic mariners wanted the wind dead on the stern, so they could run with it, sticking to the same latitude—a method called easting and westing. Columbus, Cabot, and many others had done this. Until the eighteenth century, there was no reliable way to measure longitude, which helps explain why so many could make landfalls from the Bahamas to Newfoundland and believe they were near Asia. Latitude could be fixed by the relationship to the North Star or the sun, one of the few things known in the sixteenth century about celestial navigation. Westing ships out of Bristol reached the Labrador coast and dropped down to Newfoundland. Bretons went straight to the Grand Bank and St. John's. If a ship left the harbor at La Rochelle, dropping below the two islands that protect that harbor, and then stuck to its latitude, it would make landfall at Cape Breton, where the French station of Louisbourg was established so that La Rochelle ships could easily find it.

Every spring, Europeans would arrive to fish the northern Banks and scramble for the best shore spaces for drying the catch, which were called fishing rooms. The tradition was first come, first served. The fish were dried on spruce branches, and because the scrubby northern pine forest was slow growing, the island became badly deforested from building new rooms every spring. The Europeans fished through the summer and then, before the ice hardened, tried to catch a good fall westerly to return to the European markets. Ships began leaving a caretaker behind to maintain a fishing room through the winter. This must have been one of the loneliest jobs ever created, because Newfoundland was not attracting settlers.

The single most telling fact about the island's history is that the capital, St. John's, like Petty Harbour on the other side of Cape Spear, is located on the point of land farthest from Canada and the rest of North America and closest to Europe. The entire Newfoundland economy was based on Europeans arriving, catching fish for a few months, and taking their fish back to Europe.

New England has a milder winter: ice-free harbors, a longer growing season, and arable land. Even more important in those early years, the cod moved closer to shore for winter spawning. Cod ideally spawn in a water temperature ranging from forty to forty-seven degrees Fahrenheit. Experiments show that at forty-seven degrees, eggs will hatch in ten to eleven days; at forty-three degrees in fourteen to fifteen days; at thirty-eight to thirty-nine degrees in twenty to twenty-three days. Seeking this forty-seven-degree water, cod will spawn on the coast of southern New England in the height of winter, somewhat closer to spring off of Maine, and in the summer in Newfoundland.

In the North Atlantic, farming and fishing were traditionally combined. In Iceland, which, like Newfoundland, has little arable land and a very short growing season, Icelanders were still able to combine fishing and farming, or at least shepherding. The cod run off Iceland's southern coast in the dark winter when there is little for a farmer to do. Well into this century, few Icelanders were listed as fishermen because most considered farming their primary occupation. But even if Newfoundland fishermen were to eke out a farm existence on bad land with a short season, in Newfoundland the cod ran in the summer so the farming season conflicted with the fishing season.

New England, the southernmost grounds of the Atlantic cod, had an inshore winter fishing season and an offshore summer season. It also had good farmland. While Newfoundland remained a frontier with summer fishing rooms, Massachusetts had residents who needed coopers, blacksmiths, bakers, and shipbuilders—tradesmen with families that built communities. It also became an agricultural society, settlers moving ever farther toward western Massachusetts looking for fertile land to produce goods for the prosperous coastal market. As the most flourishing American community north of Virginia, New England was perfectly positioned for trade. In cod it had a product that Europe and European colonies wanted, and because of cod it had a population with spending power that was hungry for European products. This was what built Boston.

The economies of Newfoundland and Nova Scotia, rather than growing in tandem with the economy of New England, were being drained off by it. Lacking population and internal markets, they were fishing outposts serviced by Boston. The Newfoundland catch between April and September was more than the fishing ships could hold. Trade ships, called sack ships because they carried *vin sec,* dry white French wine, came from England and took the cod to Spain, from where they returned to England with wine and other southern European products. They brought trade to England but nothing to Nova Scotia and Newfoundland. As Boston grew from its three-pointed trade, sack ships started trading between Newfoundland and Boston. The Newfoundland cod catch would be sold in Boston in exchange for Massachusetts agricultural products such as corn and cattle, which would be brought back to Newfoundland. Right up to the 1992 moratorium, many of the local Newfoundland fishermen sold their cod to Boston. So as New England grew, the northern fishing colonies remained an outer frontier with a small and stagnant population. When life in these colonies proved too hard, many moved to the prosperous sister colonies in New England.

Meanwhile, New Englanders were becoming a commercial people, independent and prosperous and resentful of monopolies. While the West Indies sugar planters were thriving on protected markets, New Englanders were growing rich on free-trade capitalism. Theirs was a cult of the individual, with commerce becoming almost the New England religion. Even the fishermen were independent entrepreneurs, working not on salary but, as they still do in most of the world, for a share of the catch. Adam Smith, the eighteenth-century economist, singled out the New England fishery for praise in his seminal work on capitalism, *The Wealth of Nations.* To Smith, the fishery was an exciting example of how an economy could flourish if individuals were given an unrestricted commercial environment.

The British Crown had never intended to grant such freedom, and now it had a colony that no longer needed it—a dangerous precedent in the midst of the empire.

THE CHOWDER AND DANIEL WEBSTER

Daniel Webster once delivered a lengthy speech on the floor of the U.S. Senate about the virtues of chowder and was considered an authority on the subject.

PARTY CHOWDER

Take a cod of ten pounds, well cleaned, leaving on the skin. Cut into pieces one and a half pounds thick, preserving the head whole. Take one and a half pounds of clear, fat salt pork, cut in thin slices. Do the same with twelve potatoes. Take the largest pot you have. Try out the pork first, then take out the pieces of pork, leaving in the drippings. Add to that three parts of water, a layer of fish, so as to cover the bottom of the pot; next a layer of potatoes, then two tablespoons of salt, 1 teaspoon of pepper, then the pork, another layer of fish, and the remainder of the potatoes.

Fill the pot with water to cover the ingredients. Put over a good fire. Let the chowder boil twenty-five minutes. When this is done have a quart of boiling milk ready, and ten hard crackers split and dipped in cold water. Add milk and crackers. Let the whole boil five minutes. The chowder is then ready to be first-rate if you have followed the directions. An onion may be added if you like the flavor.

This chowder is suitable for a large fishing party.

—Daniel Webster,
from *The New England Yankee Cookbook,*
edited by Imogene Wolcott, 1939

There are a number of other chowder recipes attributed to Daniel Webster. In his memoirs, General S. P. Lyman quotes Webster concluding one recipe: "Such a dish, smoked hot, placed before you, after a long morning spent in exhilarating sport, will make you no longer envy the gods."

Also see pages 252–56.

5: Certain Inalienable Rights

PRAY SIR, WHAT IN THE WORLD IS EQUAL TO IT? PASS BY THE OTHER PARTS, AND LOOK AT THE MANNER IN WHICH THE PEOPLE OF NEW ENGLAND HAVE OF LATE CARRIED ON THEIR FISHERIES. WHILST WE FOLLOW THEM AMONG THE TUMBLING MOUNTAINS OF ICE AND BEHOLD THEM PENETRATING INTO THE DEEPEST RECESSES OF HUDSON'S BAY AND DAVIS STRAITS, WHILST WE ARE LOOKING FOR THEM BENEATH THE ARCTIC CIRCLE, WE HEAR THAT THEY HAVE PIERCED INTO THE OPPOSITE REGION OF POLAR COLD . . . NO SEA BUT WHAT IS VEXED BY THEIR FISHERIES, NO CLIMATE THAT IS NOT WITNESS TO THEIR TOIL. NEITHER THE PERSEVERANCE OF HOLLAND, NOR THE ACTIVITY OF FRANCE NOR THE DEXTEROUS AND FIRM SAGACITY OF ENGLISH ENTERPRISE, EVER CARRIED THIS MOST PERILOUS MODE OF HEARTY INDUSTRY TO THE EXTENT TO WHICH IT HAS BEEN PUSHED BY THIS RECENT PEOPLE.

—Edmund Burke,
British House of Commons, March 1775

By the eighteenth century, cod had lifted New England from a distant colony of starving settlers to an international commercial power. Massachusetts had elevated cod from commodity to fetish. The members of the "codfish aristocracy," those who traced their family fortunes to the seventeenth-century cod fisheries, had openly worshiped the fish as the symbol of their wealth. A codfish appeared on official crests from the seal of the Plymouth Land Company and the 1776 New Hampshire State seal to the emblem of the eighteenth-century *Salem Gazette*—a shield held by two Indians with a codfish overhead. Many of the first American coins issued from 1776 to 1778 had codfish on them, and a 1755 two-penny tax stamp for the Massachusetts Bay Colony bore a codfish and the words *staple of Massachusetts.*

When the original codfish aristocrats expressed their wealth by building mansions, they decorated them with codfish. In 1743, shipowner Colonel Benjamin Pickman included in the Salem mansion he was building a staircase decorated with a gilded wooden cod on the side of each tread. The Boston Town Hall also had a gilded cod hanging from the ceiling, but the building burned down, cod and all, in 1747. After the American Revolution, a carved wooden cod was hung in the Old State House, the government building at the head of State Street in Boston, at the urging of John Rowe, who, like many of the Boston revolutionaries, was a merchant. When Massachusetts moved its legislature in 1798, the cod was moved with it. When the legislature moved again in 1895, the cod was ceremoniously lowered by the assistant doorkeeper and wrapped in an American flag, placed on a bier, and carried by three representatives in a procession escorted by the sergeant-at-arms. As they entered the new chamber, the members rose and gave a vigorous round of applause.

All of which proves that New Englanders are capable of great silliness.

At the time of this last transfer, three representatives commissioned to study the history of the carving presented a report in which they wrote extensively on the subject of the cod trade—about trading New England salt cod for salt, fruit, and wine in Europe and molasses, spices, and coffee in the West Indies. But the report, like many accounts of the New England commerce, contains no mention of one indispensable commodity in all this trade: human beings.

In the seventeenth century, the strategy for sugar production, a labor-intensive agro-industry, was to keep the manpower cost down through slavery. At harvest-time, a sugar plantation was a factory with slaves working sixteen hours or more a day—chopping cane by hand as close to the soil as possible, burning fields, hauling cane to a mill, crushing, boiling. To keep working under the tropical sun, the slaves needed salt and protein. But plantation owners did not want to waste any valuable sugar planting space on growing food for the hundreds of thousands of Africans who were being brought to each small Caribbean island. The Caribbean produced almost no food. At first slaves were fed salted beef from England, but New England colonies soon saw the opportunity for salt cod as cheap, salted nutrition.

For salt cod merchants, the great advantage of this new trade was that it was a low-end market. Cured cod can be a very demanding product. Badly split fish, the wrong weather conditions during drying, too much salt, too little salt, bad handling—a long list of factors resulted in fish that was not acceptable to the discerning Mediterranean market. The West Indies presented a growing market for the rejects, for anything that was cheap. In fact, West India was the commercial name for the lowest-quality salt cod.

In trade, it is an almost infallible natural law that a hungry low-end market, an eager dumping ground for the shoddiest work, is an irresistible market force. At first it offers an opportunity to sell off the mistakes that would otherwise have represented a loss. But producers increasingly turned to this fast, cheap, profitable product because it was easy. West India cure represented a steadily increasing percentage of the output of New England, Nova Scotia, and to a lesser degree, Newfoundland. Nova Scotia in particular specialized in a small, poor-quality, salted-and-dried product for the West Indies.

The first draw of the Caribbean for New Englanders was the salt from the Tortugas. But soon ships were coming back with not only salt but indigo, cotton, tobacco, and sugar. Only twenty-five years after the Pilgrims first landed, New Englanders were doing a triangulated trade. The best fish was always sold in Spain. Bilbao, with its wine, fruit, iron, and coal, became a major trading partner with Boston. The New Englanders then sailed to the West Indies, where some Spanish goods along with the cheapest cod were sold, and sugar, molasses, tobacco, cotton, and salt were bought. The ship would return to Boston with Mediterranean and Caribbean goods. They had made money at every stop.

Very quickly the next commercially logical step was taken. In 1645, a New England ship took pipe to the Canaries, then bought African slaves in the Cape Verde Islands and sold them in Barbados, returning to Boston with wine, sugar, salt, and tobacco. Shipments of salt cod followed, and soon salt cod, slaves, and molasses became commercially linked.

In preparing a museum devoted to the seventeenth-century commercial port of Salem, National Park Service officials carefully checked shipping documents, bracing themselves for an attack, and were relieved to have been unable to uncover any record of slaving on any Salem ship. But they should not take too much comfort in this. Aside from the fact that much slave trading was done clandestinely, the search for these records misses the important point. Regardless of how many ships actually did or did not carry slaves, or how many New England merchants did or did not buy or sell Africans, the New England merchants of the cod trade were deeply involved in slavery, not only because they supplied the plantation system but also because they facilitated the trade in Africans. In West Africa, slaves could be purchased with cured cod, and to this day there is still a West African market for salt cod and stockfish.

The French politician Alexis de Tocqueville, in his reflective 1835 study, *De la démocratie en Amérique*, wrote about an inherent contradiction in the New England character. "Nowhere was this principle of liberty more totally applied than in New England," he wrote. But then he went on to describe what he termed "the great social enigma of the United States." Freedom-loving New Englanders accepted a great deal of repression and social injustice. He described the Connecticut legal code, in which blasphemy and adultery were capital crimes, and the Boston society that crusaded against long hair. The slave trade was another example of the moral contradiction Tocqueville had observed. New England society was the great champion of individual liberty and even openly denounced slavery, all the while growing ever more affluent by providing Caribbean planters with barrels of cheap food to keep enslaved people working sixteen hours a day. By the first decade of the eighteenth century, more than 300 ships left Boston in a good year for the West Indies.

The development of a faster fishing boat, the schooner, increased production capacity of this quick cure. In 1713, the first schooner was built and launched from Eastern Point, Gloucester, by Andrew Robinson, and though there were earlier European experiments with this type of rigging, the Gloucester schooner revolutionized sailing and fishing. It was a small, sleek, two-masted vessel with fore-and-aft rigging and the ability to put a tremendous amount of canvas in topsails. The name comes from an eighteenth-century New England word, *scoon,* meaning "to skim lightly along the water." In full sail with a good breeze and a flat sea, heeling at a slight angle, the vessels did seem to scoon, and this remains one of the most elegant sights in the history of sailing. But often they were out on the Banks climbing up and tobogganing down swells as high as their masts. By re-

Woodcut of eighteenth-century Gloucester harbor. (Corbis-Bettmann)

ducing sailing time between Georges Bank and the coastline drying flakes, they increased production of West India cure.

Some of New England's best customers were the French colonies of St. Domingue (Haiti), Martinique, and Guadeloupe and the Dutch colony of Suriname (Dutch Guiana). These colonies were huge plantation economies, and the French ones were extremely profitable. After 1680, the French brought an average of 1,000 Africans to Martinique every year. Eighteenth-century St. Domingue averaged 8,000 a year. While many of the slaves were replacing others who had been worked to death, an African slave population nourished on cheap salt cod was rapidly growing.

The French fisheries were not able to satisfy this demand. The one requirement of the Caribbean market was that the saltfish be dried hard so that it could survive a tropical climate. The French lacked shore space for drying. During the eighteenth century, the limited French space in Newfoundland was whittled down to almost none. The British made their base on the eastern coast, the headlands, close to the Banks. The French fished from the south coast, Placentia Bay, where there were good ice-free harbors, the herring ran, providing bait, and New France's Gaspé Peninsula was nearby. Then, in 1713, after a fight with the British, the French agreed to leave this area and settle for access to the north coast of Newfoundland, which has been known ever since as the French shore. The area was not adjacent to other French territories, nor is it close to good fishing grounds, so it did not offer convenient drying space. After the next war, the French position became even worse.

The Seven Years War, known in the United States as the French and Indian War, was the first global conflict. In the late 1750s, France and Britain fought each other, not only in Europe but in India, the Caribbean, and North America. On September 13, 1759, New France was lost in twenty minutes when British general James Wolfe scaled the cliffs leading to the fortress at Quebec City and surprised the French garrison under General Louis de Montcalm. Montcalm, who had known previous victories against the British, made the error of leaving the fortress and meeting the surprise attack on the flat field behind the battlements. Within minutes, both generals lay dying and Quebec had fallen.

Instead of settling for battlefield victories, the British deliberated for three years over what to take from France. Some wanted to let the French keep their cod colony in North America and instead take a sugar colony as the price for peace. Guadeloupe produced more sugar than all the British West Indies combined. But the issue was never whether sugar or cod and furs was the more valuable. It was a debate about how to best hold on to North America. Given the attitudes, the economic independence, and the growing population of New England and the other lower North American colonies, Britain feared losing North America. Some argued that the French presence, an enemy at their backs, would keep the North Americans loyal to the British. But in the end, the British thought that they had better secure as much of North

America as they could. In 1763, they decided to deny France all of its North American possessions except two tiny islands off the south coast of Newfoundland, St. Pierre and Miquelon.

Ironically, when France retained Guadeloupe but lost Canada—held its slave colonies but lost its fisheries—the demand this created for West India cure in the French Caribbean led New Englanders on a direct collision course with the British Crown. The conflict went back to the Acts of Trade and Navigation, one of the foundations of the British Empire, according to which colonists were to sell their goods to England and buy their goods from England. Legally, New Englanders should not have traded directly with Spain and the Caribbean but were supposed to have sold their cod to England and then to have purchased Spanish wine and iron from England.

The British had good reasons to worry about North America. In 1677, ninety-eight years before the cause of American independence became a shooting war, the British Crown received a polite note from New Englanders accompanied by ten barrels of cranberries, two of corn mush, and 1,000 codfish. Perhaps not as bitter as ten barrels of cranberries, the enclosed note stated, "We humbly conceive that the laws of England are bounded within four seas, and do not reach America. The subjects of his majesty here being not represented in Parliament, so we have not looked at ourselves to be impeded in trade by them."

What Charles did with 1,000 codfish and all those cranberries is not known, but he did absolutely nothing about the Trade and Navigation Acts. Instead, the law was bent by the force of the marketplace. New England produced too much cod for the British market. It could not all be sold in Britain, and the British merchant fleet did not have the capacity to reexport that much cod. In spite of the Trade and Navigation Acts, the British had to allow the New Englanders to trade it.

Freed from restraint, as Adam Smith pointed out, the trade grew. By 1700, the British West Indies could not absorb all of New England's cod. Nor could it fully supply New England's rum industry, which was a by-product of the cod trade. Typical of the difference between New England and Newfoundland, Newfoundland imported Jamaican rum for local bottling, and still does, whereas New England imported molasses and built its own rum industry to sell in foreign markets. There were now three ways to buy slaves in West Africa: cash, salt cod, or Boston rum.

Massachusetts and Rhode Island rum producers were getting directly involved in the slave trade. Felton & Company, a Boston rum maker founded in the early nineteenth century, described the trade with remarkable candor in its 1936 drink guide. "Ship owners developed a cycle of trade involving cargoes of slaves to the West Indies—a cargo of Blackstrap Molasses from those islands to Boston and other New England ports—and finally the shipment of rum to Africa."

Soon the British Empire was not only too small a market for New England's cod catch but too small a molasses producer for New England's distilleries. Total British West Indies molasses production was less than two-thirds of what Rhode Island alone imported. The French colonies needed New England cod, and New England needed French molasses.

Then the British Crown, after letting New Englanders taste free trade for more than a century, decided in 1733 to regulate molasses as a key step toward reasserting its control over commerce. Instead the measure turned out to be one of the first inadvertent steps toward dismantling the British Empire.

WEST INDIA IN THE WEST INDIES

TIME SO HARD YOU CANNOT DENY
THAT EVEN SALTFISH AND RICE WE CAN HARDLY BUY.

—1940s calypso by "the Tiger" (Neville Marcano)

In Puerto Rico there was a *piropia,* a catcall to attractive women, that went *Tanto carne, y yo comiendo bacalao* (so much meat, and I'm just eating salt cod). Today meat is cheaper than salt cod, but the expression, like *piropias* themselves, persists.

Salt cod was a cheap food, mixed with other cheap foods, to make popular dishes. While it is no longer cheap, the recipes remain unchanged. Along with salt cod and roots, the most universal Caribbean salt cod dish is Salt Cod and Rice. Originally, it was a way of stretching the salt cod supply and was often made with the tail or other scraps. Sometimes a stock was prepared from the bones and the rice cooked in that, a dish known in Puerto Rico as *Mira Bacalao* (Look for the Salt Cod).

SALT COD AND RICE

This is a favourite native dish. The saltfish and rice, about a half a pound of saltfish to a pint of rice, are boiled together with the usual bit of salt pork and a little butter.

—Caroline Sullivan,
The Jamaica Cookery Book, Kingston, 1893

Also see pages 257–61.

6: A Cod War Heard 'Round the World

SALT FISH WERE STACKED ON THE WHARVES, LOOKING LIKE CORDED WOOD, MAPLE AND YELLOW BIRCH WITH THE BARK LEFT ON. I MISTOOK THEM FOR THIS AT FIRST, AND SUCH IN ONE SENSE THEY WERE,—FUEL TO MAINTAIN OUR VITAL FIRES—AN EASTERN WOOD WHICH GREW ON THE GRAND BANKS.

—Henry David Thoreau,
Cape Cod, 1851

THE ART OF TAXATION CONSISTS OF PLUCKING THE GOOSE SO AS TO OBTAIN THE MOST FEATHERS WITH THE LEAST HISSING.

—Jean-Baptiste Colbert (1619–83)

There is romance to revolution. There was to those of France, Russia, Mexico, China, Cuba. But the most romantic of revolutions, such as 1848, seem the greatest failures. The American Revolution was a remarkably successful revolution. It did not fall into chaos and violence, nor did it slide toward dictatorship. It produced no Napoleon and no institutionalized ruling party. It achieved its goals. It was also, as revolutions go, extremely unromantic. The radicals, the real revolutionaries, were middle-class Massachusetts merchants with commercial interests, and their revolution was about the right to make money.

John Adams, the most forceful of this radical Massachusetts element, did not believe in colonialism as an economic system and therefore did not believe that Americans should accept living in colonies. The American Revolution was the first great anticolonialist movement. It was about political freedom. But in the minds of its most hard-line revolutionaries, the New England radicals, the central expression of that freedom was the ability to make their own decisions about their own economy.

All revolutions are to some degree about money. During France's revolution, the comte de Mirabeau said, "In the last analysis the people will judge the Revolution by this fact alone—does it take more or less money? Are they better off? Do they have more work? And is that work better paid?" But he was not a radical in that Revolution.

Massachusetts radicals sought an economic, not a social, revolution. They were not thinking of the hungry masses and their salaries. They were thinking of the right of every man to be middle-class, to be an entrepreneur, to conduct commerce and make money. Men of no particular skill, with very little capital, had made fortunes in the cod fishery. That was the system they believed in.

These were not shallow men. Many of them, most of the important leaders—even Thomas Jefferson, a slave owner—understood that it was hypocrisy to talk about the rights of man and ignore the agony of millions of slaves. But they were not going to let the Revolution break down over this issue, as they feared it might. Throughout the century, Englishmen had predicted that the booming American colonies would try to break free from the Crown, but that, in the end, they would remain in the British Empire because of their inability to get along with each other. What the British Crown failed to understand was that the Revolutionary leaders were pragmatists focused on primary goals and that molasses, cod, and tea were not mere troubling disagreements; they were *the* issue. Virginians even called the Revolution "the Tobacco War."

England had shown some flexibility. Gloucester, though a legally recognized trade port, did not even have a customs official. The British also allowed South Carolina to trade rice for fruit, salt, and wine directly with the Mediterranean. The greatest latitude was in trade with British West Indies colonies. For Massachusetts, this trade was cod for molasses, but Connecticut traded vegetables, Maryland wheat, and Pennsylvania corn. By the 1740s, New England had as much trade with the Caribbean as it did with England. Before the English started worrying about an armed war of independence, they were worrying about a de facto independence. The colonies did not need the mother country, and both parties knew it.

Britain's first major attempt to reassert its colonial monopoly was the Molasses Act of 1733, which imposed such heavy import duties on molasses from the non-

British Caribbean that it should have virtually eliminated the trade. By making the purchase of French West Indies molasses unprofitable, the measure should have not only reduced New Englanders' markets for cod but also reduced their rum industry. It did neither, because the French were eager to work with the New Englanders in a lucrative contraband arrangement. Cod–molasses trade between New England and the French Caribbean actually grew after the Molasses Act.

The act might have been a forgotten failure had the British not tried again a generation later, with the Sugar Act of 1760, which put a six-cents-per-gallon tax on molasses. Again, New Englanders persevered through contraband. In 1764, the British tried a new tactic, actually lowering the tax on molasses, but levying new ones on sugar and on Madeira. This was intended to make colonists switch from Madeira to Port, the latter being available only through British merchants. Instead, the colonists boycotted both. Though Madeira was also traded for a middle-grade cure of cod known as the Madeira cure, rum was their drink. It was so commonplace that the word *rum* was sometimes used as a generic term for alcoholic beverages. The year of the Molasses Act, it was calculated that the consumption of rum in the American colonies averaged 3.75 U.S. gallons per person annually. In 1757, George Washington ran for the Fairfax County seat in the House of Burgesses. His campaign expenses included twenty-eight gallons of rum and fifty gallons of rum punch. There was also wine, beer, and cider. This may seem modest compared to today's campaign spending, but in 1757 Fairfax County, Virginia, had only 391 voters.

In 1764, Boston merchant John Hancock, already a known active rebel, was arrested on a charge of Madeira smuggling on his sloop, the *Liberty.* An angry Boston mob freed him. The following year, the Stamp Act for the first time charged colonists with a direct tax rather than a customs duty. As the British stepped up enforcement of trade laws, relations deteriorated. For the first time, customs agents were assigned to Gloucester, though these unfortunate officials were harassed, brutalized, and sometimes driven into hiding. In 1769, Massachusetts claimed that restraints on trade had resulted in losses for 400 vessels involved in the cod fishery.

Repeatedly, the British seemed to make the worst possible moves. Confronted with resistance to the Stamp Act, they replaced it with the Townsend Act, named for a man whose footnote in history was earned by declaring to the House of Commons—reportedly while drunk—"I dare tax America." Faced with an immediate furor over his proposed list of import taxes, he tried to back down, attempting to settle on a few less onerous items, one of which was tea.

The Boston Tea Party of 1773 illustrates the nature of the American Revolution. Here was an uprising against a tariff on an import, instigated by merchants, including John Hancock and John Rowe, in which the scions of the codfish aristocracy—dressed up as Mohawks—boarded their own ships and dumped the goods into the harbor. Similar "tea parties" followed in other ports. In New

York, evidently the Revolution had reached the proletariat, because a zealous mob dumped the goods in the Hudson before the rebels had a chance to show up in their Indian outfits.

The next British move seems even more baffling. In 1774, in response to a crisis originally provoked by the fact that the colonies produced too much surplus food, the British closed down Boston Harbor in an attempt to starve the populace until they reimbursed the Crown for damaged goods. This was not 1620, and no one was going to starve in New England, with or without imports. Marblehead supplied cod, Charleston rice, and Baltimore grain. A flock of sheep was even herded up from Connecticut.

The harshest blow to New England was to come, but communication was so slow that the colonists did not even hear of it until after the shooting had begun. The Restraining Act, effective July 12, 1775, restricted New England trade to the ports in England and barred New England fishermen from the Grand Banks. It was as though the Crown was trying to rally Massachusetts around its radicals.

During the Revolution, the American ability to produce food was the one advantage of the Continental Army. The British Army might have been better trained and more experienced, and it was certainly better dressed and equipped. But the Americans were better fed. They were also better paid, and, thanks to Boston rum, they drank better.

But there was not much cod for anyone. Newfoundlanders and Nova Scotians could no longer sell their fish in Boston. British warships kept New England fishermen from working the Grand Banks, but New England fishermen, with their fast schooners, made their waters dangerous for any pro-British ship. Gloucester schooners were outfitted with gun carriages. Ironically, the first of these armed schooners was named the *Britannia.* It was rigged with eight old cannons mounted on newly built carriages. This modest firepower was supplemented by small arms. In 1776 alone, such privateer schooners seized 342 British vessels.

In 1778, three years after the shooting began, both sides were ready to negotiate and talks began in Paris. By 1781, only three issues remained unresolved: the borderline, payment of debts to England, and fisheries. Of the three, fishing proved the most difficult.

Massachusetts insisted on fishing rights to its traditional grounds, which included the Grand Banks, the Scotian shelf, and the Gulf of St. Lawrence, all of which were off the coast of loyal British colonies. But even France, America's great ally, did not back New England. Supporting a revolution against England was one thing, but the French did not believe it was in their own interests to allow the New Englanders back into the Grand Banks. The French position was that while all nations had a right to the high seas, offshore grounds were the property of the owners of the coastline. The little islands of St. Pierre and Miquelon still allowed France to be one of the proprietors of the coastline. To this day, the French claim fishing rights in a strip of Canadian waters because of this minuscule possession.

International law of the sea was not clear on this concept. It was still widely held that the seas had no nationality. The first recognized claim on ocean territory, a three-mile limit in the North Sea, did not come into force until after the Napoleonic wars. But the New Englanders had on their side John Adams, America's most underrated founding father. It was Adams, whose face is on no currency and has inspired few monuments, who argued in the Continental Congress for complete independence from England, who won his argument by forging a Massachusetts-Virginia alliance and then bringing along the colonies in between, who chose Colonel George Washington to lead the Continental Army, who wrote "Thoughts on Government," which became a blueprint in designing the United States government, and who then plucked from their ranks young Thomas Jefferson for a protégé and assigned him to write the Declaration of Independence on the grounds that the young Virginian was a better writer than himself.

To the fury of the representatives of southern colonies, Adams had a provision written into the British negotiations declaring that fishing rights to the Banks could not be relinquished without the approval of Massachusetts. This led to one of the first North-South splits in the United States. Southerners complained that the interests of nine states were being sacrificed "to gratify the eaters and distillers of molasses" in the other four.

Adams had defended Jefferson's Declaration of Independence line by line at the Continental Congress in a nonstop two-and-a-half-day argument, while Jefferson, the reticent author, sat in silence. Among the few battles Adams lost was the one over a substantial passage against slavery, calling it "cruel war against human nature." Yet he defended the cod and molasses trade despite its slavery connection. He explained to his fellow American delegates the commercial value of the cod trade in the Mediterranean and the Caribbean. But he also argued that New England cod fishermen had proven themselves to be a superb naval force. Adams now called the New England fishery "a nursery of seamen and a source of naval power." He argued that the groundfishermen of New England were "indispensably necessary to the accomplishment and the preservation of our independence."

Many of the Americans, including Benjamin Franklin, saw fishing rights as a point they could concede. But Adams would not yield. Finally, on November 19, 1782, a year and a month after British troops surrendered at Yorktown, the British granted New England fishing rights on the Grand Banks.

But the Americans had not won access to markets. They were barred from trade with the British West Indies, a tremendous commercial loss for New England resulting in a tragic famine among slaves cut off from their protein supply. Between 1780 and 1787, 15,000 slaves died of hunger in Jamaica. In time, Nova Scotia and Newfoundland took up the slack, and their fisheries too became largely geared for low-grade West India saltfish.

* * *

During the colonial period every turn of history seemed to favor New England fisheries. But after American independence, this remarkable winning streak started to change.

The British and the Americans went to war again in 1812. Fishermen from Gloucester, Marblehead, and other New England ports, manning swift ships-of-war on a design borrowed from their schooners, did a masterful job, as Adams had predicted thirty years earlier, in "the preservation of our independence."

It fell upon John Adams's son, John Quincy Adams, to negotiate the peace in Ghent. Like the rest of his family, he was a firm advocate of New England fishing interests. Once again the issue divided the United States in a North-South split.

For New Englanders, the Paris treaty ratified in 1783 had been a huge victory. But southerners wanted to rewrite that treaty because, in order to get the British to yield on the Grand Banks, they had been given navigation rights on the Mississippi River. New Englanders insisted that their fishing rights had been signed and were not negotiable. But a Virginian, James Madison, was president, and he sided with his native South. The Ghent treaty that ended the War of 1812 denied British rights to the Mississippi but left the issue of the Banks open to further negotiation.

The Convention of 1818 reasserted some American fishing rights in the Banks, but the New England fishermen never regained what John Adams had won for them in 1782 and the issue would be a source of tension between the United States and Canada for the next 200 years.

North American cod fisheries were hurt, probably far more than either Adams would have wanted to admit, by the abolition of slavery in 1834 in the British West Indies, 1848 in the French Antilles, and 1849 in the Dutch Antilles.

After centuries of bloody slave rebellions, Europe found in the homegrown sugar beet a safer alternative to sugar colonies. Caribbeans continued to eat salt cod and to fashion drums from the barrels. In fact, now that cod no longer comes in barrels, the barrels are still made for musicians. But once the huge plantation economies ended, these little islands became very small markets.

After two centuries of dumping on the Caribbean slave market, there was little quality control in North American salt cod. This was how Thoreau found the Provincetown fishery in 1851:

> The cod in this fish-house, just out of the pickle, lay packed several feet deep, and three or four men stood on them in cowhide boots, pitching them on to the barrows with an instrument which has a single iron point. One young man, who chewed tobacco, spat on the fish repeatedly. Well, sir, thought I, when that older man sees you he will speak to you. But presently I saw the older man do the same thing.

The Mediterranean markets had constant complaints about the quality of Newfoundland cured cod. In 1895, a

RIPPING, CUTTING, AND SPLITTING

The throat of the fish should be cut as near the gills as possible. The skin between the napes (known as liver strings) should be cut on both sides to prevent tearing when opening fish for gutting. They should be ripped close by the vent on left side, all guts and liver carefully removed, and heads cut off instead of breaking off.

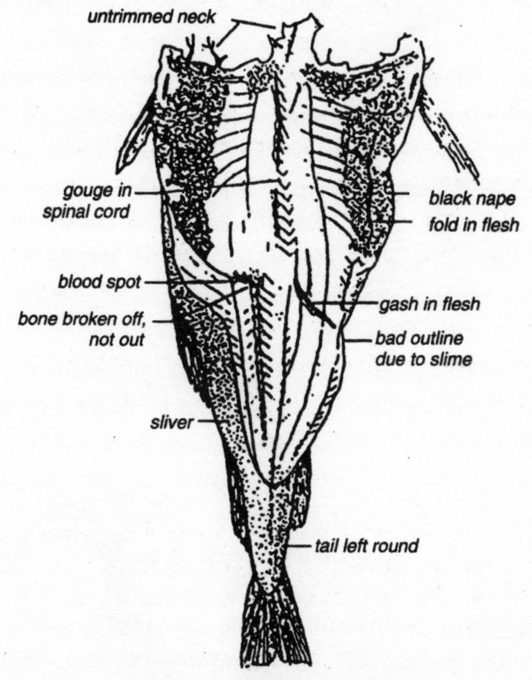

A BADLY SPLIT COD FISH

The wrong way to split a cod, from *Notes on Processing Pickled and Smoked Fish* by A. W. Fralick, senior field inspector, Maritimes Region, Canada. (Fisheries Museum of the Atlantic, Lunenburg, Nova Scotia) shipment of Labrador and Newfoundland salt cod was sent to Bilbao; the Basques, saying, "It was not liked here," shipped it on to southern Spain. In the late twentieth century, right up to the 1992 moratorium, the Canadian government was still trying to convince Newfoundland fishermen not to spear the cod in the way Thoreau had described, because it damaged the fish.

From the Middle Ages to the present, the most demanding cod market has always been the Mediterranean. These countries experienced a huge population growth in the nineteenth century: Spain's population almost doubled, and Portugal's more than doubled. Many ports grew into large urban centers, including Bilbao, Porto, Lisbon, Genoa, and Naples. Barcelona in 1900 had a population of almost one million people—most of them passionate *bacalao* consumers.

But North Americans did not succeed in this market. Though Newfoundland, Labrador, and Nova Scotia remained almost entirely dependent on fishing, there was little quality and they largely sold to Boston or the Caribbean. The one North American exception was the Gaspé, where a quality Gaspé cure was sold to the Mediterranean. Some 900 years after the Basques won the competitive edge over the Scandinavians by salting rather than just air-drying fish, the Scandinavians became competitive by perfecting salting. Norway and Denmark, which controlled Iceland and the Faroe Islands, moved aggressively into the top-quality Mediterranean markets and have remained.

Even today, with goods and people moving more freely than ever before, most salt cod eaters are attached

George Dennis's fish-curing establishment, circa 1900, east Gloucester. (Peabody Essex Museum, Salem, Massachusetts)

to the traditional cure of their region. Modern Montreal is a city of both Caribbean and Mediterranean immigrants. At the Jean Talon market in the north of the city, stores feature badly split, small dried salt cods from Nova Scotia and huge, well-prepared salt cod from the Gaspé. The Caribbeans consistently buy the Nova Scotian, while the Gaspé is sold to Portuguese and Italians.

In New England, just as the West Indies market declined, the domestic market grew. Salt cod became a staple of the Union Army, and Gloucester profited from the Civil War. But the war had also industrialized northern economies, and New England, a key player in the American industrial revolution, became much less dependent on its fisheries. The old merchant families moved their money into industry. The term *codfish aristocracy* was now used by an emerging working class to remind the establishment that they had gotten rich in lowly trade and therefore, for all their airs, were simply nouveau riche.

Their image as Revolutionary leaders faded, and, for all their aristocratic trappings, they were simply remembered as haughty people who had once made a lot of money from fish. In 1874, a Latin American revolutionary, Francisco de Miranda, visited Boston and after going to the Massachusetts State House, reported that the cod hanging there was "of natural size, made of wood, and in bad taste." Worse yet, in the 1930s, Boston's Irish-American mayor, James Michael Curley, a feisty populist who took on the Boston establishment, objected to calling them codfish aristocracy. He said the term was "an insult to fish."

A LINGERING MEMORY

In the American South, slaves modified African cooking for white people. After the Civil War, this process continued as many former slaves found jobs cooking for corporations or the railroad. "I was born in Murray County, Tennessee, in 1857, a slave. I was given the name of my master, D. J. Estes, who owned my mother's family, consisting of seven boys and two girls. I being the youngest of the family." So begins the self-published book of Rufus Estes, "formerly of the Pullman Company private car service and present chef of the subsidiary companies of the United States Steel Corporations in Chicago." Given the flaking technique in his recipe, the date, and place, the "codfish" is probably salt cod.

STEWED CODFISH

Take a piece of boiled cod, remove the skin and bones and pick into flakes. Put these in a stew pan with a little butter, salt and pepper, minced parsley and juice of a lemon. Put on the fire and when the contents of the pan are quite hot the fish is ready to serve.

—Rufus Estes,
Good Things to Eat, 1911

part two

Limits

COD—A SPECIES OF FISH TOO WELL KNOWN TO REQUIRE ANY DESCRIPTION. IT IS AMAZINGLY PROLIFIC. LEEWENHOEK COUNTED 9,384,000 EGGS IN A COD-FISH OF A MIDDLING SIZE—A NUMBER THAT WILL BAFFLE ALL THE EFFORTS OF MAN TO EXTERMINATE.

—J. Smith Homans and J. Smith Homans, Jr., editors, *Cyclopedia of Commerce and Commercial Navigation*, New York, 1858

7: A Few New Ideas Versus Nine Million Eggs

HARVEY COULD SEE THE GLIMMERING COD BELOW, SWIMMING SLOWLY IN DROVES, BITING AS STEADILY AS THEY SWAM. BANK LAW STRICTLY FORBIDS MORE THAN ONE HOOK ON ONE LINE WHEN THE DORIES ARE ON THE VIRGIN OR THE EASTERN SHOALS; BUT SO CLOSE LAY THE BOATS THAT EVEN SINGLE HOOKS SNARLED, AND HARVEY FOUND HIMSELF IN HOT ARGUMENT WITH A GENTLE, HAIRY NEWFOUNDLANDER ON ONE SIDE AND A HOWLING PORTUGUESE ON THE OTHER.

—Rudyard Kipling,
Captains Courageous, 1896

The banks are treacherous. Depths as great as eighty fathoms are found there, but also areas of fifteen or twenty fathoms and less. Occasionally, in stormy weather, rocks break the surface. Ice floes split off of Greenland and the Arctic and drift south. In 1995, a large one, ironically shaped very much like a great fish with a towering dorsal fin, drifted to the mouth of St. John's harbor. Even against the high cliffs of that well-sheltered port, it was huge—out of scale with anything around it.

At sea, it is difficult to perceive the scale of these drifting ice mountains until they are suddenly off the bow, blocking everything else from sight.

Then there is the cold. For all these centuries, men have gone out in the North Atlantic when the arctic wind froze the spray to the rigging, turning lines into one-foot-thick columns of ice, making the ships unstable from the weight of the ice on the windward side. Ice would have to be chopped off the rigging to prevent capsizing. Even with improved navigation, radar, and radio reports on ice and storm conditions, cod still has to be fished out of water that is from thirty-four to fifty degrees. Fishermen must haul lines out of these waters. Today, there are new synthetic materials to protect the hands, but until recently, fishermen wore nippers—thick rubber gloves with cotton lining. They were awkward. It was hard to mend a net with gloves on, and without them, fingers could freeze without warning in a half hour. If the fingertips start turning black, all the fisherman can do is go below to thaw them out in cold water. Warm water would cause unbearable pain. Fishing is hard on the fingers anyway, and fishermen commonly lose fingers or joints from frostbite, line snags, and machinery. Hands invariably get deep cuts that become infected. If the hands get too beaten up, permanently numb from frostbite, or have too many missing fingers, the fisherman is forced into retirement.

Fishermen like to talk about their esprit de corps, and it is true that there is a warm camaraderie, a sense of being part of an elite brotherhood. Fishermen are like

Lost in Fog by James Gayle Tyler, Russel W. Knight collection. (Peabody Essex Museum, Salem, Massachusetts)

combat veterans who feel understood only by their comrades who have survived the same battles. But fishing is a constant struggle for economic survival. Each man works for shares of the catch. Anyone who can't keep up, whether because of injury or age, is harassed out of the fishery. There are few fishermen over fifty. And because fishermen are technically self-employed and not salary earners, governments have been slow to recognize claims to social benefits for those who are out of work.

One of the worst enemies of cod fishermen, especially in the days before radio, was fog. Since cod grounds are zones where warm and cold currents meet, fog is commonplace. It can be so thick that the bow of an eighty-foot vessel is obscured from midship. A lantern on the bow cannot be detected 100 feet away. Fishermen drift in a formless gray, tooting horns and blowing whistles, hoping other craft hear them and avoid collision. But the greatest danger was for the dorymen.

From the seventeenth century to the 1930s, the common way to fish for cod and other groundfish was to go out to the Banks in a ship and then drop off small dories with two-man crews. The Portuguese, who were infamous on the Grand Banks for the harshness of their working conditions, used one-man dories. Europeans would cross the ocean in large barks built for deck space and large holds; New Englanders and Nova Scotians went out in schooners that could swiftly run back to shore to land fish; but all the dories were the same: twenty-foot deckless skiffs. The dorymen would generally use oars, and occasionally sail power, but they had to provide their own sails. Often they or their wives made them by sewing together flour sacks.

Being competitive with each other, dorymen sometimes secretively took off to grounds they had discovered. Many dorymen drowned or starved to death or died of thirst while lost in the fog, sifting through a blank sea for the mother ship. They tried to fish until their boat was filled with fish. The more fish were caught, the less seaworthy the dory. Sometimes a dory would become so overloaded that a small amount of water from a wave lapping the side was all it took for the small boat to sink straight down with fish and fishermen.

René Convenant, one of the last Breton dorymen, wrote of his father's death:

My father disappeared under 60 meters of cold Newfoundland water. Maybe he was the victim of a wave that was a little stronger than the others, against a dory loaded to the gunwales with fish. The fragile launch was filled with ice, and weighted by boots and oilskins, my father and his mate—a 22-year-old boy—sank instantly. A terrifying death without witnesses in the cottony fog that stifles all sound. Like a nightmare from which there is no awakening. . . .

"Your father, I knew him well. He was a good dory skipper." This was the only funeral oration for the missing sailor, which another sailor—Father Louis—uttered many years later when I questioned him on the tragic disappearance of my father.

To seagoing people of the North Atlantic, the hardships and bravado of dorymen were legendary. In 1876, Alfred Johnson, a Danish-born Gloucester doryman, responding to a dare, sailed his sixteen-foot boat from Gloucester to Abercastle, Wales, in fifty-eight days, the first one-man North Atlantic crossing ever recorded. Nova Scotians still recall a nineteenth-century doryman who was lost in the fog for sixteen hours before being found—the Nova Scotian survival record. But the most famous Nova Scotian doryman was Howard Blackburn, who immigrated to Gloucester. On January 23, 1883, Blackburn and his dory mate rowed away from their ship to longline halibut and became lost in a snowstorm. His mate froze to death, but Blackburn shaped his fingers around the oars so that he would still be able to row after he lost feeling in his hands. He rowed 100 miles and reached Newfoundland with the frozen corpse of his mate on the stern. Though the misadventure cost him all his fingers and most of his toes, he went to sea in sloops designed for his disability, set a thirty-nine-day, one-man Gloucester-to-Lisbon record, and even rowed the Florida coast with oars strapped to his wrists.

Not only dories were lost. Whole ships went down. John Cabot's was the first of many. The number of Gloucester fishermen lost at sea between 1830 and 1900—3,800—was 70 percent greater than all the American casualties in the War of 1812, and this from a town of about 15,000 people. On February 24, 1862, a gale swept Georges Bank, and 120 drowned in one night. In the 1870s, as schooners became shallower and carried more sails, making them even faster and more beautiful, but much more dangerous, Gloucester losses became horrendous. These shallow, loftily rigged "clipper schooners" did not stand up well in gale winds. In 1871, twenty schooners and 140 men were lost. In 1873, thirty-two vessels and 174 men were lost, 128 of them in a single gale. An easterly gale on the banks in 1879 sunk twenty-nine vessels with a loss of 249 men.

The ports that sent fleets to the Grand Banks held religious ceremonies before the beginning of what was called "the campaign." In St.-Malo, in late February, fifteen days before the Terre-Neuvas sailed, the cardinal of Rennes came to the port to say mass before the fleet. A wreath was tossed to sea to remember the fishermen who had been lost in previous campaigns.

As fishing modernized, fishermen were no longer lost in dories but were twisted in electric winches used to rapidly haul cable, slammed by trawl doors flying across the deck, crushed by rollers. On the modern trawler, being crushed in machinery is the leading cause of death but is closely followed by the more traditional fisherman's death, drowning. Ships sink at sea; men fall or are swept overboard. If a fisherman gets his foot ensnared in a rope that is rapidly paying out, he will be dragged over and drowned almost before anyone realizes he is overboard.

Fishermen do not like talking about these risks among themselves, just as Sam Lee and his Petty Harbour companions did not want to talk about the risks of falling off their open deck. But even the luckiest of fishermen have one or two stories of near-mishaps. Fishermen have the highest fatal accident rate of any type of worker in North Atlantic countries. According to a 1985 Canadian government report, 212 out of every 100,000 Canadian fishermen die on the job, compared to 118 forestry workers, 74 miners, and 32 construction workers. In 1995, 5 American workers per 100,000 died in work-related accidents, but among fishermen, more than 100 per 100,000 died. Similarly, a 1983 British study shows the death rate among British fishermen to be twenty times higher than in manufacturing.

One of the reasons for such high accident rates is that fishermen have always operated on very little sleep. If the catch is plentiful, the fishermen might go a day or two with no sleep. In the old salt fishery, once the dorymen came back on board, their catch had to be cleaned. The head was chopped off, the belly opened, the liver set aside—sometimes along with the roe, sounds, throats, and other items. Next, the cod had to be carefully split and the spine removed. (A bad split destroyed the value of the fish.) Then it had to be carefully salted. If the fishermen were lucky, they could have a few hours of sleep.

The first push to modernize fishing came from the French. In 1815, the new French government decided to subsidize the rebuilding of their fisheries, which had been devastated first by the French Revolution and then by the Napoleonic wars. Revitalizing the economy was only part of the motivation. As John Adams had once pointed out, it was far cheaper to subsidize long-distance cod fleets, which produced excellent sailors, than to maintain a well-trained standing navy. The British grudgingly began doing the same thing, but not until they had spent years complaining about the French subsidy.

The French outfitted their Terre-Neuve fleets with longlines, otherwise known as trawl lines, setlines, or bultows. Until then, the principal technique for cod fishing throughout the North Atlantic had been handlining, exactly the method Sam Lee and the other Newfoundland inshore fishermen still use. Sometimes a spreader was put on the end so that two baited hooks came off it instead of one.

Records show the British used longlines off of Iceland in 1482, and they may have been used earlier. But before the nineteenth-century French, the system had never become popular because it required an enormous quantity of bait. In Canadian waters, the French found ample herring and capelin. Though modest by contem-

Handlining. The Georges Bank cod fishery, plate 32 from *The Fisheries and Fishery Industries of the U.S.* by George Brown Goode, 1887. (Peabody Essex Museum, Salem, Massachusetts)

porary standards, these early-nineteenth-century French longlines were longer than they had ever been before. They could be as short as a half mile, or they might extend for four or five miles. About every three feet, a two-foot lanyard with a hook on the end was tied. The dory ran the line out. Caulked barrels served as buoys, which were placed at periodic distances so the line could be found. (Today the buoys are bright plastic balls with a flag on a two-foot mast over the top to make them visible from a distance.) The doryman would row along the line, hauling up, taking fish, rebaiting, and releasing.

In 1861, it was written in the *Journals of the Assembly*

in Nova Scotia, "Setline fishing, there can be little doubt, was induced by the enormous bounty of ten francs paid by the French government for every quintal [sixty-five fish] of fish caught by their fishermen. . . . The writer has been informed, incredible as it may appear, that some of these lines have as many as ten thousand hooks fastened to them." Such an operation required only a few dozen men and five dories.

As the nineteenth-century debate over longlining grew, nationalism, more than conservation, seems to have been the issue. Unfair competition from the French subsidy system angered British North America, later Canada, more than the possibility of overfishing from the technique it financed.

Distrust of new fishing techniques is endemic to fishing. Longlining had always been controversial in Iceland, as was netting when it was first used for cod in Icelandic waters in 1780. But fishermen objected to netting because they feared it would block off the fish and they would move to some other waters. The principal Scandinavian objection to longlining was that it was unfair, undemocratic. Longlining required capital to buy large quantities of bait, and those who could not afford the bait did not have the opportunity. In North America, the principal nineteenth-century argument against longlining was the same one brought up in Petty Harbour when it banned the practice in the late 1940s. As Kipling described in his novel, there were too many fishermen out on the Banks working the same grounds. If they all started using longlines, there would not be enough space and they would be constantly fouling each other's lines.

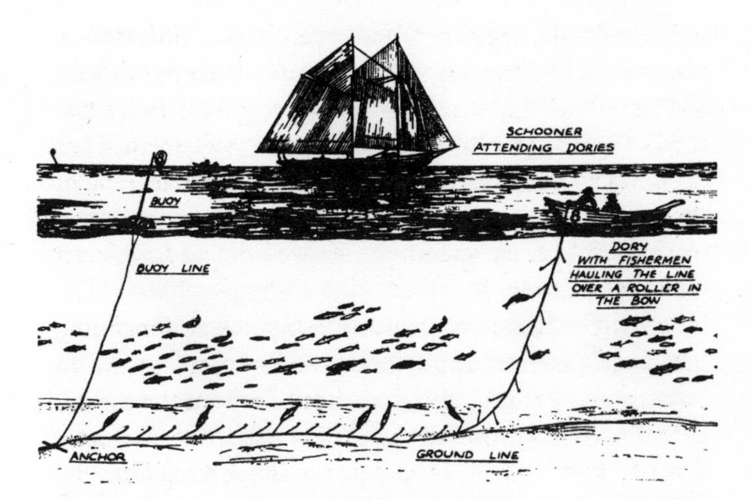

Trawl-line dory fishing (longlining), figure 4 from *Fisheries and Fishing Vessels of the Canadian Atlantic* by N. J. Thompson and J. A. Marsters. (Peabody Essex Museum, Salem, Massachusetts)

There was the rare, purely conservationist measure such as Newfoundland's 1858 law regulating the mesh size in the herring fishery. But it was difficult to think of overfishing when the catches were getting bigger every year. Catches were improving not because the stocks were plentiful but because fishing was getting more efficient. Nevertheless, as long as better fishing techniques yielded bigger catches, it did not seem that the stocks were being depleted.

The indomitable force of nature was a fashionable nineteenth-century belief. The age was marked by tremendous optimism about science. The lesson gleaned from Charles Darwin, especially as interpreted by the tremendously influential British scientific philosopher Thomas Henry Huxley, was that nature was a marvelous and determined force that held the inevitable solutions to all of life's problems. Huxley was appointed to three British fishing commissions. He played a major role in an 1862 commission, which was to examine a complaint of driftnet herring fishermen, who said that longliners were responsible for their diminishing catches. The fishermen had asked for legislation restricting longlining. But Huxley's commission declared such complaints to be unscientific and prejudicial to more "productive modes of industry." The commission also established the tradition in government of ignoring the observations of fishermen. It reported that "fishermen, as a class, are exceedingly unobservant of anything about fish which is not absolutely forced upon them by their daily avocations."

At the 1883 International Fisheries Exhibition in London, which was attended by most of the great fishing nations of the world, Huxley delivered an address explaining why overfishing was an unscientific and erroneous fear: "Any tendency to over-fishing will meet with its natural check in the diminution of the supply, . . . this check will always come into operation long before anything like permanent exhaustion has occurred."

Considering the international impact of Huxley's work in the three commissions, it is disturbing to note that he once explained his participation in these paid appointments by saying, "A man with half a dozen children always wants all the money he can lay hands on."

For the next 100 years, Huxley's influence would be reflected in Canadian government policy. An 1885 report by L. Z. Joncas in the Canadian Ministry of Agriculture stated:

> The question here arises: Would not the Canadian fisheries soon be exhausted if they were worked on a much larger scale and would it be wise to sink a larger amount of capital in their improvement?
>
> . . . As to those fishes which, like cod, mackerel, herring, etc. are the most important of our sea fishes, which form the largest quota of our fish exports and are generally called commercial fishes—with going so far as to pretend that protection would be useless to them—I say it is impossible, not merely to exhaust them, but even noticeably to lessen their number by the means now used for their capture, especially if, protecting them during their spawning season, we are contented to fish them from their feeding grounds. For the last three hundred years fishing has gone on in the Gulf of St. Lawrence and along the coast of our Maritime Provinces, and although enormous quantities of fish have been caught, there are no indications of exhaustion.

Joncas supported this assertion by referring to a British Royal Commission in which Huxley participated: "Not withstanding the enormous and continually increasing quantities of fish caught annually along the coasts of Great Britain, the English fisheries show no sign of exhaustion."

But Joncas had a political agenda for making these assertions. He believed that the Canadian government, like that of France, should become more involved in financially supporting its fishing industry. In the decades since the French had introduced longlining, the technique had become widely used in Canadian fisheries. Joncas now argued in favor of gillnetting because it did not require the great quantity of bait needed for longlining. He pointed out that gillnetting was being used by Canada's biggest competitor in the world cod market, Norway.

A gill net is a net anchored slightly above the ocean floor. It looks somewhat like a badminton net. Groundfish become caught in it and, trying to force their way through headfirst, end up being strangled at the gills. The nets are marked by buoys, and the fisherman has only to haul them up every day and remove the fish. But sometimes the nets detach from their moorings. As they drift around the ocean, they continue to catch fish until they become so weighted down that they sink to the ocean floor, where various creatures feast on the catch. When enough has been eaten, the net begins to float again, and the process continues, helped by the fact that, in the twentieth century, the gill net became almost invisible when hemp twine was replaced first by nylon and then by monofilament. Since monofilament is fairly indestructible, it is estimated that a modern "ghost net" may continue to fish on its own for as long as five years.

Joncas complained that the twenty- to thirty-foot schooners used in the Gaspé and the Prince Edward Islands were far too small to be competitive, and he recommended that the government help Canadian fishermen acquire large ships with deck space for onboard fish processing—what would one day be called "a factory ship."

The solution to Joncas's quest already existed. In midcentury, the steam engine had been invented, but fisheries were slow to seize on this machine. When they did, it would be the first new idea to dramatically change cod fishing since the discovery of North America. Soon there would be another idea: frozen food. Once these two inventions were put together, the entire nature of commercial fishing would change.

DELIGHTING IN HENCOD ROES

MR. LEOPOLD BLOOM ATE WITH RELISH THE INNER ORGANS OF BEASTS AND FOWLS. HE LIKED THICK GIBLET SOUP, NUTTY GIZZARDS, A STUFFED ROAST HEART, LIVER SLICES FRIED WITH CRUSTCRUMBS, FRIED HENCOD'S ROE.

—James Joyce, *Ulysses*, 1922

Leopold Bloom's tastes were more old-fashioned than eccentric. Until recently, cod roe was the central feature of an Irish breakfast. Most Irish today do not eat cod roe for breakfast because, though they do not seem to realize it, what is called Irish breakfast is increasingly similar to English breakfast. In the old Irish breakfast, the roe was sliced in half and fried in bacon fat or simply boiled.

BOILED ROE

It is better to buy roe raw and cook it yourself. Do not choose too large a roe; the smaller ones have a more delicate flavour.

Wrap the roe in a piece of cheesecloth and put it into a warmed salted water. Let it cook very gently—the water should just bubble and no more—for at least 30 minutes. When cooked, take it out and let it get cold. The outer membrane is taken off before using, but leave it on until you use the roe, as it keeps it moist.

—Theodora FitzGibbon, *A Taste of Ireland*, 1968

Also see pages 247–49.

8: The Last Two Ideas

SAID HE, "UPON THIS DAINTY COD
 HOW BRAVELY I SHALL SUP,"—
WHEN WHITER THAN THE TABLECLOTH,
 A GHOST CAME RISING UP!

—Thomas Hood (1799–1845),
"The Supper Superstition"

If a cod fisherman of Cabot's day could have returned to work in the year 1900, he would have been dazzled by the new inventions on shore, but once he went to sea his job would have seemed familiar. When John Cabot's voyage opened up North American waters to Europeans, the nature of cod fishing changed dramatically. Fishermen then pursued cod in much the same way for the next four centuries. True, navigation improved in the seventeenth century. The chronometer made it possible to fix a longitude in the eighteenth century. The three-masted bark with its dories was developed. New Englanders invented the schooner. Telegraph and then the trans-Atlantic cable made it possible for long-distance fleets to get news of market price trends and storm warnings. But by the beginning of the twentieth century, these inventions had only slightly changed the job of the fisherman and his ability to catch fish. Fishermen were still working the same grounds with only minor variations on the same types of gear, still in sail-powered vessels—Icelanders were still using oars—and it was still a dangerous and arduous job.

Well into the twentieth century, Lunenburg, Nova Scotia's, Grand Banks fleet stayed with sail power. "The Lunenburg cure," heavily salted on the schooners and then dried on flakes along the rocky sheltered coastline, was traded in the Caribbean. The town of Lunenburg was built on a hill running down to a sheltered harbor. On one of the upper streets stands a Presbyterian church with a huge gilded cod on its weather vane. Along the waterfront, the wooden-shingled houses are brick red, a color that originally came from mixing clay with cod-liver oil to protect the wood against the salt of the waterfront. It is the look of Nova Scotia—brick red wood, dark green pine, charcoal sea.

The Lunenburg fishery was famous for its schooners and its role in a series of Canadian-U.S. schooner races between 1886 and 1907. Then, in 1920, when the America's Cup race was canceled because of high seas, a publisher of the *Halifax Herald and Mail* who thought these sportsmen fainthearted put up a $5,000 prize and a silver cup for a fishermen's schooner race between Lunenburg and Gloucester. Fishermen, he insisted, knew how to sail schooners in rough weather. The *Gloucester Daily Times* accepted the challenge. Gorton's, the Gloucester seafood company, sponsored a schooner that beat Lunenburg's schooner twice. Then, in 1921, Lunenburg built a bigger schooner, the *Bluenose.* The competition continued until 1938, and though Gloucestermen won several races, they never took the cup away from the *Bluenose,* which can now be seen on the Canadian dime, matchbooks, and almost anywhere else eyes might fall in Maritime Canada.

Gloucester fishermen commonly worked off of schooners until World War II. Gorton's last working schooner, the *Thomas S. Gorton,* built in 1905, sailed until 1956. In 1963, Lunenburg's last fishing schooner, the *Theresa E. Connor,* sailed empty to Newfoundland because she could not find a crew in Nova Scotia to fish the Banks. No one in Newfoundland was willing to work on her either. Everything that had been done to make schooners faster had made them also more dangerous. Unable to get a crew, the *Theresa E. Connor* returned to Lunenburg, where she is still tied up as part of a maritime museum.

By then, Europeans had been using engine power in their own waters for seventy years. But because of the cost of burning coal to fuel trans-Atlantic crossings, they had been slow to convert their Grand Banks fleets. The French continued to send sailing barks with dorymen to the Grand Banks well into the 1930s, when most northern European fisheries were completely engine-powered. The last Portuguese fishing ship to work the Grand Banks without any engine power, the *Anna Maria,* went down in a storm in 1958. But it was not until the *Theresa E. Connor* was forced to tie up without a crew, 100 years after the steam engine was invented, that the age of sail in the cod fisheries finally ended.

In fishing, new technologies usually came first in Europe, where the waters had been fished longer and it was harder to catch fish than in North America. Competition for dwindling catches was the greatest incentive, and the North Sea, shared by eight affluent and fiercely competitive fishing nations, was the leading laboratory for innovation.

Originally most trawlers, ships that drag their fishing gear behind them, were longliners. But once ships had engine power, what New Englanders call a bottom dragger, which drags a net just above the ocean's floor, became the most common form of trawler. Bottom trawling was not a new idea. For centuries, the British and the Flemish on opposite sides of the Channel had caught shrimp by dragging a net along the sea floor with a wooden beam to create a wide horizontal opening at the bottom. It was pulled from shore at low tide by horses. Sail-powered draggers, known as smacks, began working in the North Sea especially after 1837, when a fishing ground called the Silver Pits, just south of the already well-fished Dogger Bank, was discovered.

These cod grounds turned the twin ports of Hull and

Grimsby on the Humber River, traditional fishing towns, into major ports. Here, steam power was first applied to fishing when steam-powered paddleboats started hauling smacks to and from the North Sea banks. Once there were steam-powered vessels, it was only a matter of time until the old beam trawl that was used on smacks got hitched to one of the new ships. First some of the paddleboats were rigged for trawling. Then, in 1881, a shipyard in Hull built a steam-powered trawler named the *Zodiac.* By the 1890s, not a single sailing trawler was left in Hull, and steam-powered trawlers were becoming commonplace in the North Sea.

The first otter trawl was built in Scotland in 1892. Instead of a beam, which would work only where the ocean floor was flat, the bottom opening of the otter trawl was maintained by a chain, which was made more mobile by metal bobbins, rollers, beneath it. The upper side of the opening was held up by floats. The net was kept wide open horizontally by "doors," heavy armored planks on either side of the net. The otter trawl is the prototype of all modern bottom draggers. By 1895, it had become the standard fishing rig of the British North Sea fleet, and very quickly the other European nations that competed in the North Sea fishery converted to otter trawls.

While the British were developing steam power to reduce time at sea, Americans were suffering staggering losses, experimenting with ever more top-rigged schooners to increase their speed. From 1880 to 1897, the years during which the British developed the North Sea steam trawler, 1,614 fishermen from Gloucester alone drowned working schooners. Yet little interest was taken in the North Sea innovations. New Englanders and Nova Scotians were stubbornly attached to their majestic, albeit deadly, schooners. Newfoundland and Labrador local fishing was inshore, where small boats trapping and handlining brought in good catches with little capital investment.

The first otter trawl was introduced to New England as an experimental loan by the U.S. Fisheries Commission in 1893 to a group of Cape Cod fishermen. But Georges Bank remained largely under sail for another three decades. Finally, by 1918, steel-hulled beam trawlers were being built in Bath, Maine, and a trawler fleet grew in Boston.

Once motor ships replaced sail and oar, fishing no longer had to be done with "passive gear"—equipment that waited for the fish. Fish could now be pursued. And since a bigger, more powerful engine could always be developed, the scale of the fishing could increase almost limitlessly.

Steam ships with otter trawls were reporting catches more than six times greater than those of sail ships. By the 1890s, fish stocks were already showing signs of depletion in the North Sea, but the primary reaction was not conservation. Instead North Sea fleets traveled farther to richer grounds off of Iceland.

The huge quantity of landings was periodically causing fish prices to crash, creating unprecedented havoc in the marketplace. In the 1920s, protests by fishermen forced the Canadian government to prohibit further expansion of the dragger fleet. The quality of cured fish was declining because once steam power made faster vessels possible, competitors vied to be the quickest to bring catches to market. In 1902, the British consul in Genoa wrote words that have proven to be prophetic: "It would be far better to return to the old system of sailer cargoes."

But technology never reverses itself. It creates new technology to confront new sets of problems. The greatest problem in commercial fishing has always been how to get the fish to market in good condition. For centuries, affluent people kept live fish in natural or man-made ponds. To keep saltwater species, they used tidal ponds where they built wooden cages. "Wet wells," watertight ship holds with holes for circulating seawater, were used as early as the sixteenth century in Holland. In the seventeenth century, British shipbuilders started including wet wells because the British did not like saltfish and there was always a greater demand for fresh fish. New Englanders also built "well smacks," ships with wet wells to transport fish to Boston and New York. But mortality was high in the crowded, sloshing, oxygen-deprived wells. Cod, ling, and other gadiforms caught in deep water could not survive in wells. The fish's sounds would fill with gas, and the disoriented fish would float to the surface and die. Fishermen tried to puncture the sounds to keep them from rising in the well.

Engines opened up new opportunities. The British experimented with wells of pumped water so that the oxygen content would be maintained. Engines also made railroads possible, which enabled landed fish to get to inland markets quickly. British ports became railroad centers.

With few people noticing, the next idea that would change North Atlantic fishing forever was being contemplated by a somewhat eccentric New Yorker, passing the winter in Labrador. Clarence Birdseye, born in Brooklyn in 1886, had dropped out of Amherst's class of 1910 because of a lack of money and, impatient with low-paying New York office jobs, had moved to Labrador with his wife, Eleanor, and their infant son to work as a fur trapper. He found that if he froze greens, they would last through the winter without losing their flavor. He filled his baby's washbasin with salted water, put cabbage in it, and exposed it to Labrador's arctic wind. The Birdseyes were the first people in Labrador to eat "fresh" vegetables all winter. This was the beginning of years of home kitchen experiments. Though the couple worked together, their son recalled Eleanor's regular irritation at finding food experiments throughout the house. He particularly remembered the fight over live pickerel in the bathtub.

Birdseye gave up trapping and moved to Washington, D.C., where he worked for the U.S. Fisheries Association. He was concerned about the practice of icing fish. In the 1820s, it had been discovered that packing fish in ice prolonged freshness. Ice, Birdseye explained, melts and becomes water, which encourages the growth of bacteria. After several more years of filling the household's

Postcard, 1910, schooner (Peabody Essex Museum, Salem, Massachusetts)

sinks and tubs with experiments, Birdseye unveiled a new technology. It required three pieces of equipment: an electric fan, a pile of ice, and a bucket of brine. Birdseye was reproducing a Labrador winter.

In 1925, he moved to Gloucester to work with fish and founded General Seafoods Company. Starting with groundfish, he also experimented with other seafood, then went on to meat, then fruits and vegetables.

It was a goose that made his fortune. The daughter of the founder of a food processing company, the Postum

Company, was yachting off Massachusetts and tied up in Gloucester. She was served a goose, which she found to be a marvelously delectable bird, and after making inquiries, discovered that it had been frozen by the local eccentric, Clarence Birdseye. She met Birdseye and learned more about the little company, which her father then bought, paying Birdseye twenty-two million dollars. Postum renamed his company General Foods, a name derived from Birdseye's General Seafoods. Birdseye believed his ideas would produce a corporate giant in the food industry comparable to General Motors or General Electric in their industries.

Birdseye improved his frozen food technology with his 1946 quick-drying process and went on to many other fields. He founded an electrical company and improved the incandescent lightbulb. In Peru, he developed a process to convert the crushed remains of sugarcane mills into paper. In his sixty-nine-year lifetime, he was awarded 250 patents.

Birdseye's introduction of freezing came at a critical moment in the cod fisheries. Americans, like the British, were increasingly demanding fresh fish, instead of cured, and the market for salt cod in the United States was steadily declining. In 1910, cured cod represented only 1 percent of fish landings in New England. But even with improved transportation, it was difficult to serve inland markets fresh fish, and so the cod market was dwindling. At the same time, the capacity of fishing fleets was greatly increasing. In 1928, the first diesel-powered trawlers were proving even more efficient than the steam-powered ones.

Because salt cod was still the major industry of Gloucester, the town was in an economic crisis. In 1923, Mayor William MacInnis met with Secretary of Commerce Herbert Hoover to discuss declining markets, and Hoover arranged a New York conference to consider ways to promote salt cod consumption in the United States. But with General Foods committed to the Birdseye freezing process, salt cod was fast vanishing from Gloucester. The same year as Hoover's New York conference, Gloucester's most established seafood company, Gorton's, had a crisis that led to the abandoning of the saltfish trade. The Italian government had purchased more than one million dollars' worth of salt cod from Gorton's. But while the order was crossing the Atlantic, Benito Mussolini came to power. When the Gorton's ship arrived, its cargo was confiscated and never paid for.

In 1921, filleting machinery was introduced to New England, and nine years later, 128 filleting plants operated in the region, selling off their waste to fish meal factories, which were also proliferating. Once freezing and filleting were put together, "fish fillets" became a leading product. Scrod, a small cod fillet, became increasingly popular. The word was used in the United States at least as early as 1849, though its origin seems to be a Dutch word, *schrode*, meaning "strip." Once filleting became industrialized, *scrod* became a household word.

But scrod was also sometimes haddock. The distinction between one groundfish and another was becoming less and less clear as fish was popularized in inland regions. Throughout the centuries, whenever cod has been popularized away from its native waters, there has been a tendency to call it simply "fish." Stockfish was originally supposed to mean dried cod but over the centuries came to be any dried gadiform. Cod and other salted gadiforms were all known in the British West Indies as saltfish. Now the same was happening with frozen fish. Consumers who previously had not been seafood eaters—some of them had never seen a saltwater species in its uncut, natural state—were buying "fish" fillets or sticks. The type of fish was seldom specified. They were thought to be cod, though increasingly they were made from haddock, until that was replaced by a boom in redfish. Today, fish sticks are usually Pacific pollock. "Fish," it seems, is whatever is left.

Fish sticks became an enormous commercial success. Fish fillets were frozen into blocks, which were then run through a saw and sliced into slabs, which were then cut into sticks. A Gorton's advertisement of the 1950s called fish sticks "the latest, greatest achievement of the seafood industry of today." It went on to say, "Thanks to fish sticks, the average American homemaker no longer considers serving fish a drudgery. Instead, she regards it as a pleasure, just as her family have come to consider fish one of their favorite foods. Easy to prepare, thrifty to serve and delicious to eat, fish sticks, it can be truthfully said, have greatly increased the demand for fish, while revolutionizing the fishing industry."

Freezing also changed the relationship of seafood companies to fishing ports. Frozen fish could be bought anywhere—wherever the fish was cheapest and most plentiful. With expanding markets, local fleets could not keep up with the needs of the companies. Gorton's and others abandoned their own trawler fleets and eventually their own ports. Between 1960 and 1970, the total U.S. production of fish sticks tripled, but Gloucester production only doubled. While business was increasing, Gloucester's market share was declining.

The most important development was that during World War II the three innovations—high-powered ships, dragging nets, and freezing fish—had come together in the huge factory ship. One of the original appeals of the steam-powered otter trawl had been that, without masts and rigging, ample deck space had been cleared for fish processing. Engine-driven ships could also have larger hulls with more storage space. Originally, the net was dragged and landed from a swinging boom on the side, a side trawler. The stern trawler, invented in the Pacific, was more stable on rough seas and could haul bigger trawls. It also provided a large, open deck space on the stern where the fish were landed. During World War II, this added space started to be used for freezing fish. By the 1950s, a time now thought of as the golden age of long-distance net trawling, cod catches were larger every year in the North Sea, off of Iceland, Norway, all of the banks, in the Gulf of St. Lawrence, and along the New England coast. Most of the world's commercial catches were increasing.

Were there any limits to how much could be caught, or was nature inexhaustible, as had been believed in the nineteenth century? Fishermen were beginning to worry. In 1949, the International Commission for the Northwest Atlantic Fisheries was formed to look for ways of controlling excessive practices.

But technology continued to focus on the goal of catching more fish. Factory ships grew to 450 feet or larger, with 4,000-ton capacity or more, powered by twin diesel engines of more than 6,000 horsepower, pulling trawls with openings large enough to swallow jumbo jets. The trawler hauled its huge net every four hours, twenty-four hours a day. Pier fishing, a technique often practiced by the Spanish fleet out of Vigo, suspended a huge trawl between two factory ships. One operated the trawl, and the other processed the fish. After the net was hauled up, the vessels switched roles and continued, so that the fishing never stopped.

The rollers along the bottom of the net were replaced by "rockhoppers," large disks that tend to hop up when they hit a rock and make it possible to drag close to a rough bottom without damaging the net. In addition, "tickler chains" stir up the bottom, creating noise and dust. Cod, and other groundfish, instinctively hide on the bottom when they sense danger, and the ticklers act like hunters beating bushes to drive birds out, sending the frightened cod out of their protective crannies and up into the nets.

The ocean floor left behind is a desert. Any fish swimming in the vast area of these nets is caught. The only control is mesh size. Fish that are smaller than the holes in the net can escape. While mandating minimum mesh sizes has become a favorite tool of regulators, fishermen often point out that once the back wall of the cod end has a good crop of fish in it, few fish of any size can escape, regardless of how big the mesh. Millions of unwanted fish—undesirable species, fish that are undersized or over quota, even fish with a low market price that week—are tossed overboard, usually dead.

For centuries, fishermen have had to study the lay of the ocean's floor and the skies. Nova Scotia fishermen used to look for what they called "cherry bottom," a type of red gravel floor favored by cod. They would drop a weighted line with a piece of tallow and bring it up to look at the color of gravel it had picked up. Or fishermen searched the horizon for a fast-forming cloud of seabirds. The air filled with furious screeches as the hungry predators dove for the sea's white churning surface to pluck baitfish—herring or capelin—from the chaos. The fishermen knew that their quarry were there: hungry open-mouthed cod and other groundfish attacking from below, forcing the desperate baitfish to flee their mid-water home. It is the food chain in all its violence, showing itself before the ultimate predator, who then knows where to cast his lines or nets.

All of these techniques are vanishing. Schools of fish are now located by sonar or by spotter aircraft, equipment developed during World War II to locate enemy submarines. Once the fish are located, the trawler can move in and clean out the area, taking not only the target catch but everything else in the area, the by-catch. As the 1950s Gorton's advertisement put it, "Thanks to these methods, fishing is no longer the hit-or-miss proposition it was 50 years ago."

NEW ENGLAND VISIT

In 1923, Evelene Spencer of the United States Bureau of Fisheries journeyed from her home in Portland, Oregon, to Gloucester, Massachusetts. She wrote about her visit in the *Portland Oregonian.* "Today 5000 [Gloucester] men sail the seas from Hatteras to the Arctic Circle." She was particularly impressed with New England salt cod cooking. "I hope Portland does not forget her coarse food and raw cabbage. But one thing she needs to be educated in is the use of saltfish products. . . . I had no idea that saltfish could be so delicious until I tasted it in Gloucester."

Then she offered two New England salt cod classics: Fish Balls and the following recipe for New England Boiled Dinner.

BOILED SALT CODFISH DINNER WITH VEGETABLES AND PORK SCRAPS

Cover the required amount of salt codfish with cold water and set on the back of the stove; when hot, pour off and cover again with cold. Change the water three or four times, allowing a half hour between. Or they may be soaked in plenty of cold water for two hours. Place in a saucepan with fresh cold water and bring to a boil, then simmer. Boiling saltfish hardens it. Add the required amount of peeled potatoes and simmer until potatoes are cooked. Gloucester says that the fish is never tough if simmered with the potatoes. Prepare beets, car-

rots and onions, boiling until tender. Cut slices of fat salt pork into small dice, place in a frying pan and fry out slowly until the fat is extracted and the "scraps" crisp.

Place the fish and potatoes in the center of the hot platter. Arrange the whole boiled beets, onions and carrots around the fish and potato. Put the pork fat and "scraps" in a hot gravy boat. When you are served some of each, the true Gloucester fashion is to proceed to cut up the vegetables and fish fine and mix it all together, then take a big ladle of pork fat and "scraps" and pour it over it all. I preferred to keep mine separate, so that I could enjoy the flavor of each.

—Evelene Spencer,
Portland *Oregonian*, 1923

9: Iceland Discovers the Finite Universe

THE TOWNS PETER OUT INTO FLAT RUSTY-BROWN LAVA-FIELDS, SCATTERED SHACKS SURROUNDED BY WIRE-FENCING, STOCKFISH DRYING ON WASHING-LINES AND A FEW WHITE HENS. FURTHER DOWN THE COAST, THE LAVA IS DOTTED WITH WHAT LOOK LIKE HUGE LAUNDRY-BASKETS; THESE ARE REALLY COMPACT HEAPS OF DRYING FISH COVERED WITH TARPAULIN.

—W. H. Auden and Louis MacNeice,
Letters from Iceland, 1967

Only a decade after reassuring the Canadians and the world that the waters around Great Britain "show no sign of exhaustion," such a thing being scientifically impossible, the British discovered that the cod stocks in the North Sea had been depleted. Finally, in 1902, seven years after the death of Huxley, the British government began to concede that there was such a thing as overfishing. Their marvelous steel-hulled bottom draggers had already moved from the North Sea to

Iceland. There they found local fishermen fishing the same way they had been when the Hanseatic League had driven the British away. Icelanders keep their traditions. They spoke, and still do, the same language as the Vikings. And they still do not have family names. Just as Eirik the Red's son, Leifur, became known as Leif Eiriksson, if a modern Icelander named Harold has a son and names him Jóhann, he will be Jóhann Haroldsson. His son will have the last name Jóhannsson; his daughter will be Jóhanndóttir.

Iceland is a lava-encrusted island, rimmed by fine, sheltered, deep-water harbors in the protective nooks of long fjords. But the fishing ports are located not in the harbors but on often harborless seaward points. Until the early decades of the twentieth century, the principal Icelandic fishing vessel was an open-decked oar-powered boat, and a harbor location deep inside a sheltered fjord would have added hours of rowing time to and from the fishing grounds. Fishermen chose a seaward point close to the fishing grounds and dragged their boat over the lava with the help of rollers made of whale ribs. Each boat might have as many as twelve oars but more commonly had four to six, with a fisherman on each oar. Usually the boat was also rigged with a small single sail, but oars were often found to be more efficient because the winds around fjords are erratic. Each oarsman operated a handline.

Yet Iceland remained primarily a fishing nation. Asked why technology did not advance, Jón Thór, historian for Iceland's Marine Research Institute, said, "After

Through the centuries, fishing vessels and gear changed little in Iceland. Above: A detail from a late-sixteenth-century illuminated manuscript, one of the *Law Books,* codifying regulations. This page was on fishing. (Árni Magnússon Institute, Reykjavík, Iceland) Below: A photograph, circa 1910, taken in Isafjördhur, a fishing community on the west coast of Iceland. (Akranes Museum, Akranes, Iceland)

1500, new ideas came very slowly." Until 1389, Iceland had been a vibrant society of literature, exploration, and creative concepts in government. But that year, Mt. Hekla erupted, shaking the entire island and covering it with long spells of darkness followed by a brutally cold winter that melted into spring floods. Then came epidemics. Then, in 1397, Iceland was transferred from Norwegian to Danish rule. The Danes ruled with indifference, declaring a trade monopoly with Iceland but pursuing almost no trade with it. In 1532, when the British were driven off, Iceland lost its last contact with the outside world.

While fish continued to be Iceland's principal export, until the second half of the nineteenth century, the quantity remained small. At the end of the eighteenth century, when New England and Newfoundland each had annual cod exports of 22,000 tons, Iceland was exporting less than 1,000 tons. Most Icelanders were farmers, but many of them, especially in the south and west, earned more fishing from February to April than farming the rest of the year. Not a tree grows on the island, except a few ornamental ones planted by landscapers in Reykjavík. So there is no fruit, nor is there grain. The English traded grain for fish in the fifteenth century. But stockfish was always the bread substitute. Pieces are torn off and spread with butter. From 1500 to 1800, every schoolchild in Iceland was given half a stockfish a day. Icelanders never acquired a taste for either fresh or salted cod. But in 1855, when the Danish agreed to lift their ban on foreign trade, Icelanders began learning how to salt cod, earning a place in quality Spanish and Portuguese markets.

Grindavík is a seaward strip of land, chosen as a fishing station in 910 because it was close to the cod grounds. In 1934, when Tómas Thorvaldsson first went to sea, fishermen still spent twenty minutes every morning dragging their boats over the lava into the sea. In the evening, it took an hour to get them back up. The people of Grindavík would go to the black lava beach in the evening and wait for their men to get home. There are still women who remember watching one of the boats capsize and get swallowed up in a huge swell.

The Icelanders of Tómas's generation grew up with little money for imports. They ate what the island produced, which was mainly every conceivable part of a codfish and a lamb. They roasted cod skin and kept cod bones until they had decomposed enough to be soft and edible. They also ate roasted sheeps' heads, particularly praising the eyeballs. Another specialty was *hákarl,* the flesh of a huge Greenland shark, hunted for the commercial value of its liver oil. The flesh, which contains cyanic acid, a lethal poison, was rendered edible by leaving it buried in the ground for weeks until it rotted.

Foreigners had never completely left Icelandic waters. In 1768, at its height, the Dutch Icelandic fleet had 160 ships fishing off of Iceland. After the French lost their North American possessions in 1763, they began fishing heavily in Icelandic waters and continued until the First World War. For Icelanders, foreign fishing vessels provided a rare and welcome contact with the outside world. But when the British returned in the 1890s, an eighty-year controversy began.

The *Hannes ráðherra,* a 1930s Icelandic trawler. Icelandic trawlers were depicted on trading cards inside packs of cigarettes sold in Iceland. The cigarettes, like the trawlers, were made in England.

There had not been a great deal of discussion over what should have been a troubling fact—that the new, large, steam-powered, steel-hulled trawlers had come to Iceland because they were so efficient as to have depleted the North Sea stocks in a decade. Overfishing had not yet become an issue. But the best cod grounds of southern Iceland are on a narrow shelf that extends only a few miles from shore. This short distance is what made an oar-driven fishery possible. Now these new British ships—huge, powerful, and made of steel—were crowding into the southern shelf, running over nets and lines of local fishermen.

The Icelanders had two opinions about this. Some

wanted all foreigners to be banned. But others thought that Iceland should get some of these monster ships itself, so it could reap the profits of its own ocean. The second argument won. It generally does. Throughout the century, when modest inshore fishermen have been confronted with powerful, efficient foreign fleets, invariably local government has decided to subsidize a competitive local fleet.

The first British-built trawler to be purchased by Iceland arrived in 1905. By 1915, the country had a fleet of twenty steel trawlers—some new, some secondhand, most of them British built.

In Grindavík, fishing continued the same way it had for centuries. The town did not *have* a harbor, and since trawlers cannot be hauled on the beach over whale ribs, this new invention meant nothing in this town, and many towns like it. But Tómas Thorvaldsson and his friends got an idea that caused the first major change in town in 1,000 years. There was a tidal pond, closed off from the sea in low tide by a narrow strip of land. Tómas and his fellow fishermen started going out with shovels and digging out the strip of land. They dug a little every day at low tide, carrying the black lava gravel away by wheelbarrow, until they had created a harbor. Tómas bought his first "deck boat," a little, low-to-the-water, steam-powered side trawler. Then he bought a bigger one. He was becoming a businessman.

Trawlers did more than increase Iceland's fishing capacity. They caused a profound change in this preindustrial society with its population of 78,000, most of whom were peasants earning little more than subsistence. The people who purchased the trawlers became Iceland's first capitalists. Cod was creating an entrepreneurial class in Iceland, the same way it had in New England in the 1640s. Reykjavík, the island's largest town, which had a population of only slightly more than 2,000 when the British trawlers first arrived, was growing into a city with a new kind of population—a working class of former farmers.

Among the developments in Iceland's improving economy was a reawakening of intellectual life. Icelandic literature, a proud tradition until the Danish takeover, was once again a creative force. The sciences also made great progress in the 1920s, and Icelandic biologists now understood that the codfish stocks had a limited capacity to reproduce.

Then the British-owned trawlers left. Just as in the age of sail, governments had seen fishermen as a low-cost way of maintaining a reserve of able-bodied seamen, they now saw their ships as a reserve of vessels easily convertible to war service. The trawlers were fast, built for rough weather and also for towing, which made them excellent minesweepers. Or, with batteries of guns mounted fore and aft, a trawler became a patrol boat. When war came in 1914, the British Admiralty commandeered most British-owned trawlers longer than 110 feet and no more than ten years old.

In an age when little attempt was made to measure fish stocks, the British Ministry of Agriculture then undertook a study of the British trawlers out of Hull and

Grimsby that had been working the Icelandic shelf. The study showed that a single trawler off of Iceland landed as many fish as three trawling the North Sea Banks. Historians now believe that Icelandic stocks would have soon been reduced to the same state as North Sea stocks had not the war provided a four-year respite. Icelandic fishermen saw their catches go up in 1917 and 1918 and then start declining again once the British returned.

After the war, British fishermen did not get back all their trawlers as promised by the Admiralty. Many were destroyed or damaged. But they were soon replaced with larger, faster, better-equipped ships. Improved landing facilities and better rail links made Aberdeen, Fleetwood, Hull, and Grimsby major ports for the British Icelandic fleet, supplying the nation through seafood companies, which, like those in New England, were becoming large corporations. But as vessels got bigger and better equipped, fishing required a greater capital investment. By 1937, every British trawler had a wireless, electricity, and an echometer—the forerunner of sonar. If getting into fishing had required the kind of capital in past centuries that it cost in the twentieth century, cod would never have built a nation of middle-class, self-made entrepreneurs in New England.

Since the industrial revolution, Great Britain had been developing an ever-increasing market for groundfish—especially cod, haddock, and plaice—because fried fish, later fish-and-chips, became the favorite dish of the urban working class. Both Iceland and Britain were fishing Icelandic water to supply this market. In addition, in the 1920s, the German fleet became a major presence in Icelandic cod grounds.

The inshore fishermen, a major sector of the economy, began protesting that their gear and grounds were being destroyed by trawlers, and among Icelanders there was an increasingly widespread sentiment that the territorial limit should be extended. But the Anglo-Danish Convention of 1901 said that the waters off of Iceland, up to three miles from shore, were open to the world, and the colony of Iceland did not have the power to change this. Instead, Icelanders built a sizable and well-trained coast guard and worked closely with the International Council for Exploration of the Seas (ICES), the accepted authority on commercial fish populations, in an attempt to monitor the size of the fish stocks. By the late 1920s, the Icelandic Coast Guard was making frequent arrests of German and British trawlers caught trespassing on inshore grounds.

The British responded by using the new wireless on their trawlers to alert each other to Coast Guard activities. Most famous were the so-called Grandmother messages of 1928. Three messages—"Grandmother is well," then "Grandmother is still well," and finally "Grandmother is beginning to feel bad"—were used to indicate when a Coast Guard vessel was leaving its harbor. Finally, in 1936, coded wireless messages were outlawed in Icelandic waters. But the Grandmother messages continued, with code systems often organized by British seafood companies.

The end of the Grandmother messages and a reprieve

for the dwindling cod stocks came in the form of World War II. Again, British distant water trawlers were all requisitioned for the war. After the Germans occupied Denmark, the Allies occupied Iceland to ensure that it, too, did not fall. Since the British now had no fishing fleet, Iceland supplied the British market and the world market. For six years, it was the only major fishing nation of northern Europe.

The British desperately needed not only food but cod-liver oil. They had a history of being great cod-liver oil enthusiasts. For centuries before it was refined for ingestion, a blackish residue from livers left in barrels was used as a balm, as it still is in West Africa. In the 1780s, British medicine decided that cod-liver oil was a remedy for rheumatism, then a catchall diagnosis for aches and pains. During the nineteenth century, it was used to treat tuberculosis, malnutrition, and other poverty-related diseases. Between the wars, cod-liver oil became a major business in Hull and was used both for livestock and humans. During World War II, the British Ministry of Food, concerned about the effect of a tightened food supply on health, provided free cod-liver oil for pregnant and breast-feeding women, children under five, and adults over forty. School nurses forcefully administered spoonfuls of the vile-tasting liquid, while adults were often given it with orange juice. All this oil came from Iceland, where it contributed to a secondary Icelandic trade that remained and prospered after the war. The British government, believing that the oil had produced the healthiest children England had ever seen, despite bombings and rationing, continued the program until 1971. It was finally discontinued because people refused to take the oil. Icelanders, however, still take it, as do many Americans.

During the war, fish prices reached record highs and were paid to Iceland in dollars, which facilitated doing business with the American troops on the island. Icelanders built American housing and military bases and began importing large quantities of American goods.

By the time the war ended, Iceland was a changed country. Not least among the changes, in 1944 it had negotiated full independence from Denmark. Now it was free to negotiate its own relations with the rest of the world. Because of cod, it had moved in one generation from a fifteenth-century colonial society to a modern postwar nation. W. H. Auden, who had spent much time there in the 1930s, returned in 1964 and was astounded by the transformation. He ran into one of his former guides, now a schoolmaster, and asked him what life had been like for Icelanders during the war. "We made money," replied the schoolmaster.

A NEW DANISH YEAR

Danes will order a whole cod weeks in advance to ensure getting a fresh one for New Year's Eve. In a Scandinavian winter funk, Ann Bierlich, a television film editor in Copenhagen, wrote that Fresh Cod with Mustard Sauce, a traditional New Year's dish, "is the only good thing about this country." She was very specific about the proper way to cook fresh cod. Fish mustard, a strong, grainy mustard, is made fresh and sold in Danish markets.

FRESH COD WITH MUSTARD SAUCE

Allow ½ kilogram per person and cut the cod in approximately 3-centimeter-thick slices. Take the biggest pot you have. Put in enough cold water to cover the fish. Use a lot of salt, a good handful of coarse salt, so salty that you think it is the North Sea. Put the pieces of cod into the cold water, put the stove on low heat, and warm the water so it doesn't boil, but quivers. Let the pieces of cod stay in the water a few minutes while it is still quivering. Don't cook it to death! The cod must be firm and white.

Mustard sauce—2 tablespoons of butter are melted, and 2 tablespoons flour added, and then 1 liter of milk. The sauce must be thick. Finally add a generous amount of fish mustard! (A little splash of dill vinegar will give an extra zing.)

The dish is eaten with boiled finger potatoes, chopped hardboiled eggs, chopped boiled red beets, fresh horseradish. What to drink with it? Ice cold beer and schnapps, of course!

—Ann Bierlich, Copenhagen, 1996

10: Three Wars to Close the Open Sea

LIFE IS SALTFISH.

—Halldór Laxness, Reykjavík, 1930s

When World War II ended, the fish stocks in the European North Atlantic, after six years with little fishing, were at a level that has never been seen since. There were tremendous catches on the Icelandic shelf, on the North Sea banks, in the Barents Sea, in the Channel, and in the Irish box, as all of the waters surrounding Ireland have come to be known. Like the old days on the North American banks, huge cod were commonplace. But the principal fishing

Postcard, 1910–15, printed in Iceland for French fishermen to send back home.

nations came back with ever bigger, faster, and more efficient trawlers.

With the creation of the new independent Icelandic state in 1944, the Anglo-Danish Convention of 1901, with its three-mile limit, was nullified. After five and a half centuries of indifferent colonial administration, Icelanders were determined to build a modern society through management of their one natural resource, cod grounds. By 1955, when Halldór Laxness won the Nobel prize for literature, the harsh life of prewar Iceland that he described in his novels was already becoming a faded memory. A major step in this nation building took place in April 1950. After the required two-year notice was up, Iceland annulled its old treaty and extended its territorial limit to four miles off its shoreline. A modest claim by contemporary standards, this was a bold move in 1950, when the concept that the seas belonged to everyone was a widely held principle of international law.

The three-mile limit had first been established in 1822, with the North Sea Fisheries Convention, signed in the Hague by France, Germany, the Netherlands, Denmark, and Britain. Ironically, the British were the great advocates of the three-mile zone and defended it with some of the same tactics that they termed unlawful harassment when later employed by Icelanders against them. When a British-American treaty giving American fishermen access to Canadian waters expired in 1866, the Americans were charged fees to fish in the three-mile zone. But by the 1930s, the principle of any limit was considered highly questionable to most Western nations.

Then, in 1945, because the United States wanted to protect its offshore oil production, a new concept in international law appeared. President Harry Truman issued a proclamation stating that the United States had the right to control mineral resources on its continental shelf. No one had ever owned a continental shelf. The banks did not belong to the United States and Canada. England did not own its shelf. No one owned the North Sea. Since cod and most other commercial fish are mostly found on continental shelves, the implications for fishing were enormous. Furthermore, on the same day, Truman issued another proclamation: "In view of the pressing need for conservation and protection of fishery resources, the Government of the United States regards it as proper to establish conservation zones in those areas of the high seas contiguous to the coasts." The measure was in response to a prewar dispute with Japan, whose fishermen had been catching Alaskan salmon from the sea before the fish could return to their spawning rivers.

The proclamation immediately resonated in newly nationalistic postwar Latin America, where many countries started claiming their continental shelf. Europeans—especially the British—fiercely objected, but their arguments were weakened by pressure from their own American possessions. While protesting the principle, Britain claimed a piece of the shelf for the Bahamas. By 1950, there was some international backing for Iceland's four-mile zone, especially since most international fishing was farther offshore.

But for that very reason, many Icelanders thought the new law was too modest. After 1954, Icelandic cod catches began to fall dramatically. The same was true for ocean perch, or redfish, which was an increasingly important commercial catch. In all, the groundfish catch in Iceland between 1954 and 1957 dropped 16 percent. The argument for extending the limit was further bolstered by the fact that catches of haddock and plaice, which swim closer to shore and therefore were protected by the four-mile limit, had increased during those same years. In 1958, Iceland extended its territorial limit to twelve miles.

While Icelanders were celebrating, the British government sent a formal letter of protest, which stated, "Claims to exercise exclusive jurisdiction in relation to fishing in areas outside the normal limits of territorial waters are wholly unwarranted under international law." The statement went on to say that "Her Majesty's Government find it difficult to believe that the Icelandic Government would use force against British fishing vessels in order to secure compliance with a unilateral Decree which parties of the Government Coalition propose to issue without regard for international law." Once again, the British had underestimated the zeal of a people first embracing nationhood.

And so began what the British press labeled "the Cod Wars." There were three, though there was never a declaration of war nor a single death. The lack of casualties can only be attributed to a great deal of luck on both sides.

France, Belgium, Denmark, Germany, the Netherlands, and Spain all backed the British position, which is about as close as western Europe has ever come to a united front. To them, the Icelanders were "harassing" lawful shipping beyond their territorial limits. By the withdrawal deadline, August 30, 1958, all foreign vessels, except British trawlers, had left the twelve-mile limit. They were now accompanied by British warships. The Icelandic Coast Guard spotted thirty-seven ships of the Royal Navy and 7,000 men, though Admiralty records show that the force was even larger.

The destroyers and frigates, manned by World War II combat veterans, were capable of speeds of up to thirty knots. The Icelandic Coast Guard had seven ships, of which the largest and most advanced could do seventeen knots. The ships had one gun each, and the sailors were either policemen or civilians—men with no combat experience. But they were all experienced seamen with an intimate knowledge of Icelandic waters. During the two and a half years of the first Cod War, the Coast Guard managed to arrest only one British trawler out of Grimsby that had ventured within the old four-mile limit where the Royal Navy was not patrolling. But with all of this tense maneuvering, the trawlers were getting very little fishing done. The Navy had the trawlers fishing within defendable thirty-mile-long rectangular boxes—a good military move, but abysmal for catches.

Based on its ambassador's reports, the British government believed that Icelanders were divided on the issue of extending their exclusive fishing zone. Opposition parties had voiced disagreement, but mostly on the timing of implementation. When the British realized their mistake, they began negotiating—in Reykjavík, London, and Paris. After five months, in February 1961, Britain finally recognized the twelve-mile limit, and Iceland declared its intention to look into further expansion. The Icelandic government almost fell over its agreement to give the British a three-year adjustment period.

Ten years later, the two countries repeated the same exercise. In March 1971, Iceland declared that effective September 1, 1972, it was extending its limit to fifty miles. Britain and West Germany, now partners in the European Economic Community, vehemently protested and asked the International Court of Justice to intercede in what they claimed was a violation of international law. Iceland said it did not recognize the court's jurisdiction because the action was on Iceland's continental shelf and therefore was not an international issue. Before the International Court could reach a decision, the second Cod War was fought and settled.

This second war was shorter and more dangerous. On the one side, the Icelandic Coast Guard was better prepared, with faster ships. On the other side, the Royal Navy was backed up by the West Germans, who were not allowed to have a military but provided "supply and protection" vessels. In the first Cod War, the only way the Coast Guard could have stopped a British trawler was to fire on it. Not only were they badly outgunned, but, as the British had suggested at the time, Icelanders did not really want to shoot and kill British seamen. But by 1958, Icelandic engineers had secretly developed a Cod War weapon, and the following year, they had armed all seven Coast Guard vessels with it. When negotiations in the first Cod War began, the Icelanders decided not to use this weapon and managed to keep it under wraps for more than ten years until the second Cod War.

In the second Cod War, a Coast Guard vessel would approach a foreign trawler to inform the captain that he was in violation of Icelandic law and should move outside the fifty-mile limit. If the captain did not agree, the Coast Guard ship would come about and, cruising at a right angle to the trawler, cross its path astern with the Coast Guard's secret weapon pulled behind—a "trawl wire cutter." In reality, the new weapon applied the old technology of minesweeping to fishing. One of the device's four prongs would ensnare a trawl cable and cut it, letting loose a net worth $5,000 and whatever catch might be in it. A trawler without a trawl had nothing to do but go home. During the one-year conflict, eighty-four trawlers—sixty-nine British and fifteen German—lost their nets. To protect their nets, trawlers started fishing in pairs, one working and the other guarding the stern. But since one of every two trawlers was no longer fishing, the fishing fleet was reduced to half its normal capacity. Also, several trawlers collided from following too close in rough seas. Gale winds of sixty miles per hour are common in these waters.

After the effectiveness of the trawl wire cutter was demonstrated, the second Cod War degenerated into dodgem cars on the high seas. Trawlers attempted to prevent Coast Guard vessels from cutting their trawl by ramming them. But the Coast Guard vessels could also ram, and their reenforced hulls, built for icebreaking, made them particularly effective at this. The British were reluctant to send in the Royal Navy again because Iceland and Britain were now allies in the integrated forces of NATO. Instead, they sent four large fast tugboats, which, according to Icelanders, were to ram Coast Guard vessels. According to the British, the tugs were under orders not to ram but to position themselves so that Coast Guard vessels could not cut trawler cables. Whatever their intention, the tugs did ram a few Icelandic vessels. On March 18, 1973, an Icelandic gunboat fired live shells across the bow of a British tug, and on May 26, an Icelandic shell blew a hole in the hull of a British trawler. The British trawlers withdrew to fifty miles and refused to come back until the Royal Navy protected them. Seven British war frigates moved in, and once again the trawlers were ordered to fish in boxes. If a Coast Guard ship entered a box, frigates, tugs, and trawlers would all attempt to ram it until it sank. This led to regular collisions on the high seas but, miraculously, no loss of lives or vessels. After ships collided, both damaged vessels would limp back to their home ports.

The Icelandic government was shockingly tough. It refused to allow injured or sick British seamen into Iceland unless they arrived on their vessel, which would have meant surrendering the trawler. It blocked British NATO planes from Icelandic air traffic control and even threatened to cut diplomatic relations. Unlike Britain, Iceland depended on fishing for its entire economy; fishing was the miracle that had lifted its people from the Middle Ages to affluence. Despite a history of warm feelings between the two nations and a close alliance, Iceland was not going to yield on its only resource. NATO, concerned about this conflict within its ranks in the middle of the Cold War, began pressuring Britain to back down. In the end, Britain recognized the fifty-mile zone in exchange for limited arrangements for smaller British trawlers.

One of the great changes in the postwar world was that small nations were making themselves heard as never before through international forums, most notably the United Nations. The idea of expanding sovereignty into the ocean was catching on. In a 1973 meeting of the UN

Seabed Committee, thirty-four nations, mostly Latin American, African, and Asian, endorsed the concept of a 200-mile zone. The only northern European nations to endorse this were Iceland and Norway, both of which were fast becoming leaders in the cod trade.

In 1974, Icelandic cod stocks appeared to be in trouble again, in spite of the fifty-mile limit. The percentage of large cod in the catch had declined dramatically. Icelandic biologists claimed that a decade earlier eighteen-year-old cods were commonplace, whereas by 1974 it was rare to find a cod older than twelve. This meant the reproductive capacity of the stock was greatly reduced. Even British scientists agreed with these findings.

One more time, on October 15, 1975, citing diminishing cod stocks and the need for conservation measures, Iceland extended its limit, this time to 200 miles. And once again, all foreign trawlers sailed outside the new zone except the British and, this time, the West Germans. They were going to do it all over again.

It was to be the shortest and meanest of the three wars. In one incident in December 1975, a British tug reported that an Icelandic Coast Guard vessel fired two shots, neither of which hit. In five months, there were thirty-five ramming incidents as the Icelandic Coast Guard cut forty-six British and nine German trawls. Both sides were becoming practiced at the arcane skill of friendly naval battles. British foreign secretary James Callaghan told the home press, "Both sides in the conflict are showing valor, but there is no need for anyone to show their virility."

Negotiations were also intense. "The Icelanders are, by any standards, very difficult to deal with," reported the London *Financial Times.* Iceland was not going to compromise. At one point, the nation actually severed diplomatic relations with Britain. But NATO continued to pursue talks. Jón Jónsson, longtime director of Iceland's Marine Research Institute and one of the negotiators for the third Cod War, said, "Between scientists it was a very friendly cod war. The English are our best enemies." He recalled that a British negotiator once jokingly suggested that no trawls be cut the following Thursday because there was a program he wanted to watch on television. Jónsson fondly recalled the good travel tips he received for his upcoming vacation, when he and his wife toured Cornwall.

Although the British did not think the future of their entire economy was at risk, they had much at stake. Leaders of the trawler industry and chip shop guilds, known as fish fryer associations, warned that the entire British fishing industry was about to collapse. The great cod ports of Hull, Grimsby, and Fleetwood were in precipitous decline. The merchants who bought from the trawlers and sold wholesale were dependent on Icelandic cod. Between the end of the second Cod War and 1976, the number of wholesale merchants in Hull had dropped from 250 to 87. Britain's allies repeatedly suggested that the problem could be solved by British consumers abandoning their beloved cod for other species. The West Germans negotiated a truce with Iceland in which they were given redfish quotas as compensation for being barred from catching Icelandic cod. The West German government told the British that the entire problem could be solved if British households would learn to eat redfish and pollock. The European Economic Community pointed out that blue whiting was abundant off of Scotland. "If the British could be brought to eat it, the whole cod war would become unnecessary," said the *Financial Times* in May 1976. But the British wanted to eat cod, not whiting or pollock, and they detested redfish.

Jónsson described the London negotiating sessions as "heated discussion but on a gentlemanly level. But the British fought a losing battle, and I was surprised at how shortsighted they were. All the world was going to 200 miles. I said to the British minister, 'I am quite sure you are going to 200 miles in a few years, and then we will be able to advise you on how to do it.' And they did go to 200 miles, though they never asked our advice."

Indeed, the entire European Economic Community was about to declare a 200-mile zone. The British government insisted on a 100-mile exclusive British zone in its own waters. In February 1976, the EEC embarrassed Britain, in the middle of its negotiations with Iceland, by openly rejecting the British demand and simply establishing a European 200-mile zone.

Tómas Thorvaldsson was fifty-seven years old when the 200-mile zone changed his life. He had already been through many changes and become the prosperous executive of a trawler company with its own processing plant. Now he was also part of government: a board member and, after twenty-two years, president of a state bureaucracy that controlled all fish exports. He remembered the old days and liked to visit the black crescent of shore, the lava beach where they used to drag the boats to the water. Solemnly he would say, "On this spot men have gone to sea for more than 1,000 years."

But the old Iceland seemed unimaginable to his children. Even the food was different. Young people did not eat stockfish; they went to a bakery and bought bread. Their diet, with the exception of lamb and fish, was imported and expensive. One day each winter, when the arctic night is almost twenty-four hours, and thoughts turn suicidal, a feast is held at costs unimaginable in the old days, in which the old Icelanders eat *hákarl,* sheep's head, ram's testicles, and other foods from their past.

Physical traces of past centuries vanished almost completely. If a building survived from the 1930s, it was considered historic. One-third of the town of Heimaey, a fishing port on a small island off the southern coast, was buried in a 1973 volcanic eruption. A sign in the new lava fields proclaims that buried some feet below is Iceland's oldest Kiwanis Club, built in 1924.

Most towns had a population of 2,000 or less. The few people drove cars on needlessly wide, well-paved streets. There were no crowds, no signs of poverty, nothing old and absolutely no dirt. Just a treeless immaculate plain of new houses, metal or concrete, freshly painted in colors that are true but not hot and mirror nothing in nature. Most towns looked like a sales lot for new, oversized mobile homes. "These are the new houses built for the young people after we were through digging a harbor by hand," said Tómas.

The harbor in Grindavík was dug out a little more each year. The men of the town are still dredging. It is now home port to fifty fishing vessels, ranging from small, two-man boats with a converted stern rigged for dragging, to large, modern bottom draggers.

After Iceland's 200-mile zone gained acceptance in 1976, most nations declared their own 200-mile zones. Some 90 percent of the world's known fishing grounds fell within 200 miles of the coast of at least one nation. Fishermen now had to work not so much with the laws of nature as the laws of man. Their primary task was no longer to catch as many fish as possible but to catch as many as were allowed. The fisherman had long been a skilled navigator, seaman, biologist, meteorologist, mechanic, weaver, and mender. Now he also had to learn, like a good civil servant, how to work the regulations, sidestep their pitfalls, and sail through their loopholes. He became skilled at this as well. Fishermen rarely consider regulation their responsibility. As they see it, that is the duty of government—to make the rules—and it is their duty to navigate through them. If the stocks are not conserved, government mismanagement is to blame.

If the zone is used to exclude foreigners, as most are, the nation only has to regulate its own fishermen. That was seen by Icelanders as the key to effective management. Jóhann Sigurjónsson, deputy director of the Marine Research Institute, said, "It's enough to have your own people to watch. You don't want an Olympic fishery like the North Sea. Everybody tries to take as much as they can as fast as they can." The European Community tried to solve this through its regulatory bureaucracy, the Common Fishing Policy, but that only created a new, complex set of nation-by-nation regulations for fishermen to work on.

The Icelandic government realized that it would have to curb the capacity of its own fleet. It required larger mesh on trawls. But the fishermen compensated by buying more trawlers. Then the government restricted the size of the fleet and the number of days at sea; the fishermen responded by buying larger, more efficient gear. The cod stocks continued to decline. In 1984, the government introduced quotas on species per vessel per season. This was a controversial and often wasteful system. A groundfish hauled up from fifty fathoms is killed by the change in pressure. But if it is a cod and the cod quota has been used up, it is thrown overboard. Or if the price of cod is low that week and cod happen to come in the haddock or plaice net, the fishermen will throw them overboard because they do not want to use up their cod quota when they are not getting a good price.

In 1995, a system was initiated to restrict the total cod catch to a maximum of 25 percent of the estimated stock. That also had loopholes. But with each measure, there was less and less resistance. When Icelanders see cod stocks diminishing, they think about returning to the Middle Ages—earthen huts, metal shacks, the buried shark and burned sheep heads. National politicians, fishermen, trawler owners, and seafood companies became increasingly cooperative with the scientists at the Marine Research Institute. Their greatest opponents were local politicians trying to bring something home for the district.

Before the 200-mile zone, Tómas Thorvaldsson had never thought about overfishing, only about how to catch more fish. But now he had to limit his fishing capacity. "Thinking about fishing less was very difficult for the mind," he said. He showed an empty dormitory that until 1990 had housed up to fifty-two workers from other parts of Iceland. They would come to process 2,000 tons of saltfish a year. In recent years, Tómas had processed only 300 to 400 tons a year. Higher prices, fewer fish, and fewer fishermen was the new formula of the Iceland fishery. Although the sector drove the economy, the government had already reduced the number of fishermen to only 5 percent of the workforce.

Looking around the walls of his office, where he had hung photographs of every vessel he had ever owned, Tómas pointed to that low-to-the-water little steamship, his first decked boat, and said, "Maybe we should go back to this."

BABES IN ICELAND

In late January and February, during the spawning season, it is a tradition in Iceland to eat cod roe stuffed with the fish's liver. Like most traditional Icelandic food, this dish is not popular with the young and affluent generation.

STUFFED COD ROE

Cut the side of the roe and turn it inside out. Put the liver inside. Cook in boiling water for a few minutes. Sometimes, instead of liver, I make a pudding with mashed cod, minced onions, flour, and egg because the babies don't like liver.

—Úlfar Eysteinsson,
Thrír Frakkar restaurant, Reykjavík, 1996

Also see pages 247–49.

part three
The Last Hunters

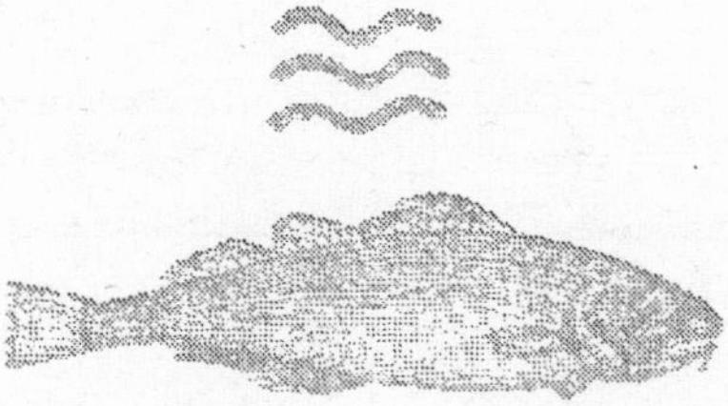

IT'S NO FISH YE'RE BUYING: IT'S MEN'S LIVES.
(FISHMONGER TO A CUSTOMER HAGGLING OVER THE PRICE OF A HADDOCK.)

—Sir Walter Scott,
The Antiquary, 1816

11: Requiem for the Grand Banks

NOW A LULLING LIFT
AND FALL—
RED STARS—A SEVERED COD-

HEAD BETWEEN TWO
GREEN STONES—LIFTING
FALLING

—William Carlos Williams,
"The Cod Head," 1932

Inevitably, Iceland and Newfoundland are compared. They are both North Atlantic islands and roughly the same size, though Newfoundland's half million inhabitants are twice as many as Iceland's. The poor quality of the land and shortness of the growing season make agriculture unprofitable on both islands. Historically, both economies have been entirely based on fishing, mostly cod. On both islands, the local fishermen operated small boats inshore while foreigners fished the rich offshore grounds. Both remained underdeveloped colonies until after World War II.

But that is when everything becomes different. While Iceland was severing its ties with Denmark to become an independent republic, Newfoundland was severing its ties to Britain and becoming a province of Canada. Once it became a province of a large wealthy nation, Newfoundlanders no longer needed to depend on their fishery for survival. Canada would make up the shortfalls. By the 1990s, the Canadian government was spending three dollars on fisheries for every one dollar those fisheries earned.

Newfoundland, Britain's oldest colony, had been a self-governing colony until the Great Depression. At that time, saltfish was failing to support fishermen, and a British-appointed commission took over. But under the Commission of Government, an unemployed fisherman received an allowance of only six cents a day. Though Newfoundlanders had always resisted the idea of being swallowed up by Canada, this appeared to be the only option left. In 1948, the British supervised a referendum in which Newfoundlanders voted by a narrow margin to become the tenth province of Canada. But once part of Canada, Newfoundland had a large and distant government that was not accustomed to thinking of fishing as a top priority. The Canadian foreign trade bureaucracy was far more interested in wheat and industrial products. It viewed the local salt cod fishery as an economic failure and tried to develop the Newfoundland economy with light industry, most of which also failed, because it could not compete with mainland industry.

Breton fishing fleet leaving for the Banks, *The Graphic*, October 17, 1891.

But once the 200-mile limit was established in 1977, the Canadian government saw a chance to make fishing a viable economic base for Newfoundland. First, though, it needed to settle its border with the United States and drive off the Europeans. Then it would have a truly exclusive zone.

The Spanish and the Portuguese, who regarded it as their right to fish these grounds because they had been doing so for 500 years, were shocked. The 200-mile limit had been a particular blow to Spain because, though its people had the highest per capita fish consumption of any Western country, almost no good fishing grounds could be found within 200 miles of the Spanish coastline. After Franco's death in 1975, every sector of the Spanish economy attracted investments and was being modernized. But there were few prospects for a modern Spanish fishing fleet. The Canadians and the Americans were throwing Spanish ships off of their banks, and the French and the British were pressuring the European Community bureaucracy to exclude them from European waters. While fishing was one of the principal objections of the French and British to letting Spain into their community, it was also one of the incentives Spain had for joining. With a larger fleet than any European Community country, Spain saw its quota reduced by the EC every year. By 1983, 1,000 Spanish vessels shared 234 licenses in European waters. Below the hilltop town of Vigo, in Spain's northwestern region of Galicia, was a fleet of modern trawlers that increasingly had nowhere to go.

The Portuguese fleet, called the White Fleet because during World War II it had painted its ships white to remind German submarines of Portuguese neutrality, had few places to fish either. The one place left to the Iberians was a corner of Grand Bank and all of the Flemish Cap, a historic cod bank, both of which were beyond 200 miles and therefore in international water. Both the Spanish and Portuguese were taking significant quantities of cod from this area until 1986, when the Canadians decided to deny foreign vessels fishing the outer Banks use of St. John's for supplies and repairs. The French still had St. Pierre and Miquelon, but the Iberians would have been stranded far from home with no port.

The other issue for Canada was the U.S. border. While Georges Bank, the richest prize on the shelf, is off the coast of New England, much of it is also within the 200-mile range of Nova Scotia. The fight over Georges Bank may not have become a true cod war in the European tradition, but a few gunshots were exchanged between New England and Canadian fishermen—probably the only shooting between Canadians and Americans since the French and Indian War. Under international arbitration, Canada was granted the northeast corner of the Bank, and the rest became U.S. territorial water. For the first time in history, Canada and the United States now exclusively owned the cod banks off their coasts.

The 200-mile limit was not seen in Canada, the United States, or anywhere else as a conservation measure, but rather as a protectionist measure for the national fisheries. While the U.S. government was providing low-interest loans and other incentives to modernize a New England fleet on Georges Bank, Canada was investing in a Grand Banks fleet. To build up this modern industry, the seafood companies, near bankruptcy from mismanagement, an overvalued Canadian dollar, and competition from Iceland, had to be rescued. Under a government bailout plan, the Newfoundland seafood companies were merged into a conglomerate called Fishery Products International, and government funds were used to resuscitate the Nova Scotian company called National Sea Products. By the late 1980s, both companies were huge and prospering. FPI had even managed to buy back the government shares. By then, the Canadian dollar was weak against the U.S. dollar, and Newfoundland

and Nova Scotia cod were commanding excellent prices in the Boston market.

Ten years after the 200-mile limit had been declared, the year after the ports were closed to foreign vessels, the Canadian government could, and did, claim that it had taken possession of its banks and turned the Atlantic fishery around into a profitable sector of the economy. There was a significant increase in the number of fishermen and the number of fish-processing-plant workers. The seafood companies crewed huge trawlers with new fishermen, many of whom were fish-plant workers, since much of the work on board a modern trawler is fish processing. Sam Lee of Petty Harbour recalled with a slight sneer, "One fellow I grew up with—his father had a store. He worked in plants. Before long he was an experienced deep-sea fisherman."

But while the new, offshore all-Canadian fishery was prospering, the inshore fishermen found their catches dropping off. They suspected the reason was that the offshore draggers were taking so many cod that the fish did not have a chance to migrate inshore to spawn. The inshore fishermen complained to the regulatory agency, the Department of Fisheries and Oceans, but the government had invested in offshore fishing, not inshore, and its political priority was to make its investment a success story. As the inshore stocks dwindled, the debate became increasingly acrimonious. On the one side were the inshore fishermen; on the other side were the fishermen's union, the trawler workers, the seafood companies, and the government. Cabot Martin, a Newfoundland lawyer who took up pro bono the cause of the inshore fishermen, said, "The whole unfair thing about Sam Lee fighting National Sea is that Sam didn't have any money."

"We founded the Newfoundland Inshore Fisheries Association because nobody was listening to the fishermen. We were complaining to the wind," said Sam Lee. "It was not just for inshore fishermen. Anyone who cared about what was happening could join."

The small local fish plant had chronic bankruptcies, and catch would spoil before the Petty Harbour fishermen could find another way to get it to Boston. Finally, the fishermen took over the plant as a cooperative, and since the government was interested in seafood companies, they were able to borrow money for improvements. By keeping the fish in pens, they could keep the cod alive until they had arranged the market. Cabot Martin, whose original interest was fish farming, showed them that by feeding the cod capelin, herring, and mackerel, they would double the weight of the catch, and much of the added weight was not in length but in thickness, increasing the fish's per-pound value. The cod started to resemble the thicker stock of Georges Bank. But as time went on, it was getting harder to get cod of any size to put in the pen. Even bait fish for feed were becoming scarce.

In 1989, faced with government indifference, Martin and the Inshore Fisheries Association decided to sue the government in the hope of getting an injunction against bottom dragging. They charged that the Department of Fisheries and Oceans was not following environmental assessments. The court ruled against an injunction, saying it would have a negative impact on the economy and force National Sea's plant in St. John's to close down for several months a year.

Martin has since observed environmental campaigns against whale and seal hunting, such as those by Greenpeace, and he regrets having gone to court at all. "McDonald's was the biggest buyer [of the draggers' catch]. We should have had a campaign against McDonald's. We weren't very sophisticated," he said.

That the government was not listening to the inshore fishermen is an understatement. The government was euphoric about Atlantic cod stocks and the future of the fisheries. Catches were rising, and fishermen who could not meet their quotas of redfish were given supplemental quotas of cod to make up the difference. A government task force under Senator Michael Kirby was charged with assessing the future of Atlantic fisheries. Much of its report was devoted to finding new markets for all the fish that was going to be caught by the new Canadian groundfishing fleet.

Canadians have never been fish eaters. Even Newfoundlanders and Nova Scotians do not eat large quantities of fish. This is also true of Americans, including New Englanders. But the U.S. population is so large that there is always a potential for expansion. According to the Kirby report, Americans consume 233 pounds (105.8 kilos) per person of red meat in a year and only 4 pounds (1.8 kilos) of groundfish. The report estimated that over the next five years, Canadian groundfish catches would increase by 50 percent, and if somehow American per capita groundfish consumption could be increased by a mere .1 percent, the U.S. market could absorb all of the Canadian surplus.

In reality, catches were increasing not from an abundance of fish but because the efficiency of a modern trawler fleet made it possible to locate the sectors with remaining cod populations and systematically clean them out. In retrospect, this seems obvious, but it must be remembered that during Newfoundland's long history of fishing, the migratory cod periodically disappeared from certain sectors only to reappear in others. Almost every year that records were kept, there were some areas of Newfoundland or Labrador where the cod stocks had nearly vanished. In some years, only one area failed. The years 1857 and 1874 were notable because there were no failing grounds. In 1868, almost all sectors experienced a failure in the stocks. But they would always show up somewhere the following year. Despite cries of alarm, these failures had never resulted in the disappearance of cod but had only been caused by temporary shifts in migratory patterns, perhaps in response to temperature changes. In the 1980s and early 1990s, the Canadian government assumed that Newfoundland waters were again experiencing this well-known phenomenon. Ralph Mayo, a marine biologist for the U.S. National Marine Fisheries Service who studies Georges Bank from the Woods Hole, Massachusetts, laboratory, calls this "the perception problem." He said, "You see some cod and assume this is the tip of the iceberg. But it could be the whole iceberg."

Furthermore, the Kirby report was still being influenced by Huxley's teaching about the resilience of indestructible nature. The idea itself seems to have more resilience than nature, and every year one or two books are still published on this idea. As with the sixteenth-century belief in a westward passage to Asia, the theory cannot be killed by mere experience.

In 1989, Fisheries minister John Crosbie, son and grandson of influential St. John's fishing merchants, stood in St. John's Radisson Hotel and tried to put to rest suspicions that the fisheries would soon have to be closed. In July 1992, he returned to the same hotel to announce just that—a moratorium on fishing the northern cod stock, putting 30,000 fishermen out of work. Sam Lee and other inshore fishermen, who had been calling for the moratorium on trawling for years, waited outside. When Crosbie refused to see them, Lee, normally a pleasant, good-humored man, began angrily pounding on the door.

In January 1994, a new minister, Brian Tobin, announced an extension of the moratorium. All the Atlantic cod fisheries in Canada were to be closed except for one in southwestern Nova Scotia, and strict quotas were placed on other ground species. Canadian cod was not yet biologically extinct, but it was commercially extinct—so rare that it could no longer be considered commercially viable. Just three years short of the 500-year anniversary of the reports of Cabot's men scooping up cod in baskets, it was over. Fishermen had caught them all.

* * *

The fish-processing plants, which had been used to justify the court's rejection of the inshore fishermen's case, were closed down anyway. The two giant companies, FPI and National Sea Products, scaled down their operations and began processing cod from Iceland and Norway. National Sea Products used a 250-employee plant in Arnold's Cove, a typical Newfoundland fishing town like Petty Harbour, built in the crevice of a bay, out on the water on top of stilts. The Newfoundland government, trying to resettle the inhabitants of the small islands off of Newfoundland in less remote places, had moved villagers to Arnold's Cove, where there were jobs at the National Sea Products plant. The plant bought Russian cod, beheaded and frozen, from the Norwegians. In Arnold's Cove it was partially thawed, filleted, and refrozen.

The community-owned processing plant in Petty Harbour wanted to do the same thing but did not have the capital. "We looked into Russian cod to keep our plant going. But it was way out of our league," said Sam Lee. Instead, they kept the plant open as a school. But they owe the government more than one million Canadian dollars in interest on money borrowed to buy freezing equipment. "We were paying it back until the moratorium. The government doesn't want the plant, so we will be able to keep it. No one wants it. Fish will come back and it will be back in operation, but by then with interest, it will be a two-million-dollar debt."

The government also had to ban the blackback fishery, because fishermen going after this bottom-feeding flat fish seemed to get suspiciously large quantities of cod in their nets. It was legal to take these cod as a by-catch, but it began to look as though fishermen were targeting the by-catch. Some fished for lumpfish, which Lee completely objected to as wasteful since the roe is taken and the rest of the fish is discarded. Some of the inshore fishermen have turned to crabbing, which has been very profitable, and others to lobstering. There have been experiments with fishing whelks for export. But to groundfishermen, these were lesser forms of fishing. Most fishermen just collected the package and waited.

St. John's, the oldest city in North America, was built on a deepwater harbor sheltered by high majestic cliffs. The town, with its brightly painted late-nineteenth-century wooden houses, overlooks the harbor from a steep hill. Despite the ornateness of the Victorian architecture, there is a frontierlike rough-hewn charm to the town. The waterfront used to be crowded with stores selling supplies to the European fleets, whose ships would line the piers at the bottom of town. Portuguese and Spaniards would play soccer in town and drink wine with crusty bread. They were all gone now. The waterfront was filled with bars, restaurants, and shops for tourists.

The constant theme of tourism was cod. White strips of peanut butter–filled hard candy were called codfish bones. Little wooden models of trawlers were sold. Bars offered an initiation to foreigners called "being screeched in." This was a holdover from the cod and molasses trade, its meaning now lost. The tourist would down a shot of Screech, a Jamaican rum bottled in Newfoundland, and then would have to kiss a codfish—usually a stuffed one. There were no other codfish except frozen Russian fillets or the occasional catch from the Sentinel Fishery.

Meanwhile, oil has been found on the Grand Banks. A decade earlier, when oil was found on Georges Bank, fishermen had played an important role in blocking the oil companies. In Newfoundland, fishermen have already expressed concern about the effect on fish of the oil companies' seismic soundings, but without an income, they do not represent a very strong lobby. "They say it [sounding] doesn't affect fish, but they're lying," said Lee.

Everyone talks of "when the cod comes back." Lee said the fish plant would reopen when the cod came back. Tom Osbourne, procurement manager for National Sea Products in Arnold's Cove, said, "Local fish will come back before too much longer, and we will go back to processing local fish. It will be king again someday. It will regain the U.S. market."

Cabot Martin believes the cod will be back. "I'd rather there were fish to fight about. It's all coming back. They will try. They will want to start dragging again. We will have to fight them again."

But nature may have different plans.

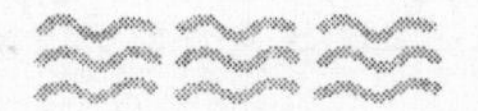

SUNDAY IN NEWFOUNDLAND

SALTED COD SOUNDS

2 lbs cod sounds
4 strips salt pork
shelots or onions

Put about 2 lb. of salt cod sounds in water & let stand overnight, then drain off water. Put in a saucepan and cook for about 10 minutes. Drain. Fry pork, cut up shelots or onions, then cut sounds in small pieces and fry altogether. Add a little water if necessary.

This recipe was used some 80 years ago, and often, for Sunday evening meal with home made bread and butter. It was enough for the family and very tasty and delicious. Today, mashed potatoes, french frys, whole potatoes with green peas could be served with this dish.

—Winnifred Green, Hants Harbor, Newfoundland, from *Fat-back & Molasses: A Collection of Favourite Old Recipes from Newfoundland & Labrador*, edited by Ivan F. Jesperson, St. John's, 1974

Also see page 249.

12: The Dangerous Waters of Nature's Resilience

WHAT WE GAIN IN HAKE, WE LOSE IN HERRING.

—English proverb

COD COMING BACK, FISHERMEN SAY
MINISTER UNDER PRESSURE TO END
MORATORIUMS IN WATERS OFF NEWFOUNDLAND

—front-page headline,
Toronto *Globe & Mail*, October 5, 1996

Newfoundlanders debated over when "the cod was coming back." Few dared ask if. Or what happens to the ocean if they don't come back? Or whether commercial fishing was going to continue at all. The position that the cod would return was most candidly argued by Sam Lee: "They're coming back because they have to."

Scientists are not as certain. Ralph Mayo of the National Marine Fisheries Service laboratory in Woods

Hole, Massachusetts, pointed out that there is no known formula to predict how many fish—or, in scientific language, what size biomass—are required to regenerate a population or how many years that might take. Both miracles and disasters occur in nature. In 1922, for unknown natural reasons, the Icelandic cod stock produced so many juveniles that, in spite of British and German trawlers, Iceland had a healthy-size stock for ten years. "There are lots of natural variables. All it takes is a huge winter storm to wash the larvae off the bank and away," Mayo said. There is only one known calculation: "When you get to zero, it will produce zero." How much above zero still produces zero is not known.

Fueling optimism is the fact that decimated cod stocks have been restored fairly quickly in other countries. In 1989, the Norwegian government realized its cod stocks were in a serious decline. It severely restricted the fishery, putting many fishermen, fish-plant workers, and boat builders out of business and drastically reducing the size of its fleet. The northern Finnmark region had an unprecedented 23 percent unemployment rate. But because the government instituted these measures while the stock was still commercially viable, while there were still some large spawners left, the cod population stabilized and started increasing after a few years. Peter Gati of the Norwegian Seafood Export Council said of the Canadian situation, "I guess politicians didn't have the courage to put people out of business." But in Norway, courage combined with good fortune and a fast-growing cod stock. When the cod stocks in the Barents Sea were mea-

Agúst Ólafsson, a deckhand aboard the *Ver,* poses with a cod for the ship's chef, Gudbjartotur Asgeirsson, circa 1925. Asgeirsson, who cooked on Icelandic trawlers between 1915 and 1940, often took photographs. (National Museum of Iceland, Reykjavík)

sured in the fall of 1992, government planners were as surprised as they had been in 1989. After the two most productive years ever recorded in this stock, the cod population was healthy again.

In 1994 the Canadian government estimated that its moratorium would last until the end of the century. Since then, politicians have tried to speed up the process. But in Canada, if everything else went well, about fifteen years would be needed to restore the population. A healthy population requires some large old spawners, and such fish in the northern stock are about fifteen years old. It is hard to imagine the Canadians holding off that long, going a generation without cod fishing. As George Rose, the fishery scientist at St. John's Memorial University, suggested, political pressure makes it almost impossible to maintain a moratorium until cod stocks have returned to historic levels. Rose, who had been a leading voice calling for the moratorium, said, "I am not optimistic that we will ever let it come back to what it was. If we get 300,000, there will be unbearable pressure to fish it."

Periodically a "food fishery" is announced. For one weekend, locals are allowed to fish cod for their own consumption. After such weekends, cod suddenly becomes available, sold off the back of pickup trucks. And yet local politicians complain that the food fisheries are too short. The mayor of Lewisporte said that some people worked on weekends and she "wanted everyone to have a chance."

In the October 1996 *Globe & Mail* article, Fisheries minister Fred Mifflin said that the Sentinel fishermen were reporting increased number and size. "The fish are fatter, they are healthier, so we know for sure that the decline has ceased." This does not at all correspond with the findings of Sam Lee and his Petty Harbour colleagues, but they are only six out of 400 Sentinel fishermen in Newfoundland. A closer look at Mifflin's data reveals that these good results were in southern Newfoundland, where waters are warmer and growth is faster. In fact, the cod there are a completely separate population from the northern stock, which inhabit the waters off the rest of Newfoundland, Labrador, and the Banks. This again illustrates what Ralph Mayo calls "the perception problem."

Weeks before Mifflin's statement, Rose said, "We found 15,000 cod in the South Bay, and everyone said the cod are back. Hold on! Ten years ago, the biomass, the population, was 1.2 million."

Some propose to give nature a hand. When the Norwegian fishery was in crisis, the government there invested heavily in experimental cod farming. Once the wild stocks returned, the Norwegians immediately lost interest in farming because it was more expensive. But fish farmers had been technically successful in transferring wild juveniles to pens and feeding them until they were thick and large. The cod were even trained to come at feeding time. "Juvenile raising is where our wild fishing is headed," said the Norwegian Seafood Export Council's Peter Gati. The flesh of the Norwegian farmed cod was extremely white because they were "purged," starved for several days before going to market, just as lobsters commonly are. Another advantage was that they could be brought to market live. This had also been key to Cabot Martin's plan in Petty Harbour.

Although farming cod is a new field and salmon farming is firmly established, Martin claims cod would be far easier to farm. Salmon have a delicate scale structure and are prone to infections, whereas cod tolerate handling and are disease resistant. Also, salmon do not like to be crowded into a pen, whereas cod have a herding social structure.

Fish farming—everything from salmon to mussel—is becoming a bigger industry every year. Farming starts out well enough. After the Petty Harbor experiment, Martin set up several pens, fattening the cod with mackerel, herring, and capelin. This probably produced excellent fish, but at the time of the moratorium he had gone out of business with a debt of one million Canadian dollars. Commercially successful fish farms reduce operating costs by feeding pellets of pressed fish meal rather than wild bait fish. In the case of salmon, they are also fed artificial coloring to give them the pink tint they acquire in the wild from eating crustaceans. Gastronomically, a wild salmon and a farmed salmon have as much in common as a side of wild boar has with pork chops.

Not only gastronomes but also scientists have deep concerns about fish farming. Pen-reared cod have a phenomenal growth rate. They are much bigger at a given age than wild northern stock. Cod doubles its size in a year anyway, but a hatchery cod can quadruple its size in the same period. Since size determines fecundity, pen raising and releasing would appear to be a way to rebuild stocks. But this is a dangerous business.

The idea of releasing farmed fish into a wild stock frightens scientists because man does not select fish in the same way nature does. If a cod was not disease resistant, did not know how to avoid predators, lacked hunting or food-gathering skills, had a faulty thermometer and so did not produce the antifreeze protein or the ability to detect a change in water temperature that signals the moment to move inshore for spawning, this cod would not survive in the wild. But it would survive in a pen, and if it had other characteristics that were particularly well suited for farm life, the defective fish would flourish and possibly even dominate. If it then reproduced with a wild fish, it would pass its "bad genes" to their offspring.

Christopher Taggart, fisheries oceanographer at Dalhousie University in Halifax, compared farmed fish to purebred dogs and thoroughbred horses: "Most purebred dogs carry genetic defects like bad hips. Thoroughbred horses break a leg if you look at them. It is a byproduct of selecting. Try to produce a dog with thick fluffy fur that is a good swimmer and it ends up to also have bad hips. If the dog bred in the wild, you would produce a wolf population with bad hips."

The genetic consequence of fish farming are still unknown. The assumption—the hope—for fish that live their entire life cycle in pens is that they never escape into the wild to mingle with the species. But this accident has happened. Worse, some hatcheries produce young for the purpose of releasing and enhancing the wild stock.

New England salmon hatcheries released so many fry into the wild that by 1996, only an estimated 500 Atlantic salmon in New England still had the diverse genetic characteristics of the wild species.

The central issue to the survival of a species is how

to maintain its diversity—the wide range of genetic characteristics that gives a species the ability to adapt to the many challenges of life in the world. Scientists have no way of knowing, but can only hope, that the tiny reduced population of surviving northern stock carry the full range of traits once presented in the gene pool of a population of many millions. Taggart argued that to preserve genetic diversity, assuming it is still there, farming "should be kept as natural as possible—an almost wild hatchery. We know that spawning places are not chosen by chance. Choose places conducive to good survivorship of juveniles and conducive to keeping the group together." That is how a wild cod chooses her spawning ground.

Overfishing is a growing global problem. About 60 percent of the fish types tracked by the Food and Agriculture Organization of the United Nations (FAO) are categorized as fully exploited, overexploited, or depleted. The U.S. Atlantic coast has witnessed a dramatic decline in the bluefin tuna population, though Gloucester fishermen refute this on the grounds that they still have good catches. Mid-Atlantic swordfish stocks are diminishing. Conch and redfish are vanishing from the Caribbean. Red snapper, which is a by-catch of shrimp, is in danger of commercial extinction in the Gulf of Mexico. Peru is losing its anchovy population. Pollock is vanishing from Russia's Sea of Okhotsk. With 90 percent of the world's fishing grounds now closed off by 200-mile exclusion zones, fishermen have been searching greater depths for new species. Little is known about the ecology of these depths, but since they often have very cold water, reproduction is probably very slow. Orange roughy was introduced to the world markets after implementation of the 200-mile zone and immediately gained such popularity that five tons an hour were being hauled up from the depths near New Zealand. In 1995, the catch nearly vanished.

The collapse of the Soviet Union destabilized many fishing agreements. Russia has become a major cod fisher, and cod has become almost the equivalent of cash in the Russian Barents Sea fishery. The reason the Canadians have been buying Russian cod processed in Norway is that Russia has been flooding the Norwegian market.

With the Atlantic long overworked by Europeans, the action has been switching to the Pacific, where not only are there large Japanese, Russian, American, and Korean fleets, but the Chinese, who do not have a history of international cooperation, have been notably enlarging their fishing capacity.

Replacing the Atlantic with Pacific fisheries is an old idea. Pacific cod was one of the reasons the United States bought Alaska from the Russians in 1867. But since the major markets were far away along the Atlantic, the Pacific cod did not have the same success as the Atlantic cod. Nevertheless, in 1890, a half million Pacific cod were landed. An 1897 book by an American scholar, James Davie Butler, suggested that with the alternative of a Pacific cod fishery, the only remaining bone of contention between the United States and Canada, cod fisheries, would be less important, and the way would now be cleared for "eventual union with Canada."

But the Pacific cod is a different fish, its flesh less prized. It does not migrate, and it does not appear to live more than twelve years. More important, the catch has never measured up to that of its Atlantic cousin. Instead, walleye pollock has become the prize of the northern Pacific, "the cod of our times," as a Gorton's employee put it, and that fish is becoming so overfished that not only its stocks but its predators, sea lions and several species of seabirds, have dramatically declined since the mid-1970s.

Marine ecology is complex and tightly interwoven. When large factory ships in the North Sea overfish sand eels and other small fish that are ground into fish meal for heating fuel in Denmark, not only cod but seabirds go hungry. In 1986, seal herds ranged south in the North Sea and ate the coastal fish off of Norway because they were famished from the overfishing of capelin. Fishermen were calling for a seal hunt to save the North Sea fisheries from the seal. In 1995, both Norway and Canada rescinded their ban on seal hunting because the populations were growing and they eat cod.

In the late 1950s, Canada's seal hunt had become a target of environmentalists when high prices for seal pelts and a huge herd had drawn packs of ruthless, unskilled amateur hunters with helicopters to the Newfoundland and Labrador coasts. In 1964, the anger of animal lovers throughout the world was stirred by a film made by Artek, a Montreal film company, that depicted a seal being skinned alive. The international protest did not abate when it was revealed that the skinner had been paid by the film company and that two of the other "hunters" turned out to be part of the film crew. In 1983, after intensive pressure from environmental groups culminated with a European Community boycott of seal products, Canada finally banned seal hunting, a traditional activity in Newfoundland and Labrador.

Not surprisingly, the 1995 reopening of the seal hunt met with national and international condemnation from environmentalist and animal-rights groups. The seal defenders claimed that there was no scientific basis for the seal hunt. Some even denied that seals eat cod. Before protecting the seal became a cause célèbre, everyone in Newfoundland knew that seals ate cod. The familiar label of the leading Newfoundland soft-drink company, G. H. Gaden, is a seal on an ice floe with the words *keep cool.* But in the less politically correct nineteenth century, a cod was in the seal's mouth.

According to the Canadian government, the seal-hunting ban caused the harp seal population to double to 4.8 million, and if the ban had not been rescinded, it would be expected to be at 6 million by 2000. Seals eat enormous quantities of fish and are particularly disliked by fishermen because they are wasteful. Like the average North American consumer, gray, harbor, and harp seals do not like to deal with fish bones. They tear into the soft belly of the cod and leave most of the rest. "Seals don't have to eat a lot of cod to have a big impact," said George Rose. "It doesn't mean we have to declare war on the seal. But we have to control the seal population." One

Canadian journalist, recalling Brigitte Bardot's 1977 campaign in which she posed on an ice floe with a stuffed baby seal, suggested that the French actress pose hugging a codfish.

Given the interdependence of species, the fundamental question is whether other species—not just the seals but the phytoplankton, the zooplankton, the capelin, the seabirds, and the whales—will wait fifteen years for cod to return. Nature may have even less patience than politicians. "Whatever will work is going to work. It will not necessarily come out the same way," said Rose. If the species that were eaten by cod become plentiful because the cod are not there to prey on them, other species may move in, and if the intruders are successful, there might not be enough food to support a large cod population again. Some biologists worry that rays, skates, and dogfish, which are small sharks, may already be moving in.

In addition, an unwanted relative has already shown up: the arctic cod *(Boreogadus saida)*. This may not be bad for the marine ecology, but it is very bad news for fishermen. Arctic cod are about eight inches long and until now have been deemed of little commercial value. Because they are a much smaller fish, the adults do not compete with the Atlantic cod for food, but the young do. Even worse, arctic cod eat Atlantic cod eggs and larvae.

The arctic cod is one of several more northerly species that seem to have expanded their range south into Newfoundland and Labrador waters at about the time the cod vanished. The other two, snow crab and shrimp, have been very profitable. Traditionally in Newfoundland, crabbing has been of a lower social order and fishermen have resisted it, but the Asian market for snow crab is extremely lucrative and several Newfoundlanders became wealthy in the mid-1990s from it. The landed value of snow crab in 1995 was the highest in dollars of any catch in the history of Newfoundland fisheries.

Scientists are not certain why any of these three species moved south. It may have been because cod, which eat shrimp and crab, were no longer there, but that would not explain the presence of the arctic cod. It may also have been that the water was colder in those years.

But Rose, who goes to sea to study the northern stock, said, "Fishermen are seeing many strange things that are a sign things are not right." The cod have been reaching sexual maturity younger and smaller. Undersized four-year-olds are spawning. This is not surprising. When a species is in danger of extinction, it often starts reaching sexual maturity earlier. Nature remains focused on survival. But Rose also said that cod were seen spawning in water temperatures of minus one degree Celsius. Cod are supposed to move to warmer water for spawning. Fishermen keep reporting aberrations, such as fish in an area where they have never been seen before, or at different depths, or a different temperature, or at a different time of year.

Perhaps even more disturbing, Rose's studies have concluded that the northern stock has stopped migrating. The stock had normally followed a 500-mile seasonal migration, but Rose believes that after 1992, the survivors came inshore and stayed. He does not know the reason for this but speculates that the bigger, older fish were the leaders and are no longer there to lead. It is also possible that cod migrate because they need food and space for spawning. With the population so reduced, this is no longer necessary.

Whatever steps are taken, one of the greatest obstacles to restoring cod stocks off of Newfoundland is an almost pathological collective denial of what has happened. Newfoundlanders seem prepared to believe anything other than that they have killed off nature's bounty. One Canadian journalist published an article pointing out that the cod disappeared from Newfoundland at about the same time that stocks started rebuilding in Norway. Clearly the northern stock had packed up and migrated to Norway.

Man wants to see nature and evolution as separate from human activities. There is the natural world, and there is man. But man also belongs to the natural world. If he is a ferocious predator, that too is a part of evolution. If cod and haddock and other species cannot survive because man kills them, something more adaptable will take their place. Nature, the ultimate pragmatist, doggedly searches for something that works. But as the cockroach demonstrates, what works best in nature does not always appeal to us.

THE PARIS DEBUT OF FRESH SALT COD

COD IS SO BEAUTIFUL, THE WAY THE FLESH UNFOLDS IN WHITE LEAVES.

—Alain Senderens

A star since he opened his first Paris restaurant when he was only twenty-nine, Alain Senderens is a culinary genius with a knack for marketing and a curious and contemplative intellect. Few people have thought as much about food as Senderens.

In 1972, as the much-talked-about young chef of his new Paris restaurant, L'Archestrate, one of his many iconoclastic ideas was to serve fresh cod, *cabillaud.* It had never before been offered in a top-rated Paris restaurant. Like salt cod, most great Paris chefs have their roots in southern regions—Senderens is from the southwest. Salt cod, *morue,* had slowly made its way up from peasant food in the south to become an honored French tradition. But not fresh cod.

He created a fresh cod recipe and offered it on the menu as *Cabillaud Rôti,* Roasted Cod. No one bought it. So he removed the word *cabillaud* and substituted *morue fraîche,*" fresh salt cod. It was a hit. This was the recipe.

MORUE FRÂICHE RÔTI
(ROASTED FRESH SALT COD)

4 220-gram pieces of cod with the skin
Eggplant caviar
750 grams diced green pepper
1.5 kilos diced red pepper
1.5 kilos chopped mushrooms
500 grams crushed tomatoes
20 chopped shallots
10 chopped garlic cloves
20 chopped anchovy fillets
10 eggplants roasted for 4 hours at 60 degrees [Celsius; 140 degrees Fahrenheit]
3 zucchinis cut into julienne strips

1. Mix olive oil, shallots, garlic, and anchovies, add red and green pepper and tomato. Cook until the liquid evaporates. Then add the roasted and crushed eggplant and mushrooms. Cook a few minutes and set aside.

2. Fry the zucchini juliennes.

3. Pan fry the cod skin down and finish cooking in the oven.

4. Arrange it on the plate.

—Alain Senderens, chef, Paris

13: Bracing for the Spanish Armada

CONFINED AS THE LIMITS OF FIELD LANE ARE, IT HAS ITS BARBER, ITS COFFEE-SHOP, ITS BEER-SHOP, AND ITS FRIED-FISH WAREHOUSE.

—Charles Dickens,
Oliver Twist, 1837–1838

In a Britain that has seen many of its most treasured traditions under siege, an issue debated with increasing frequency is the survival of true fish-and-chips. "We can foresee a time when we won't get any chunky pieces," said Maureen Whitehead, who with her husband owns the popular Polsloe Bridge fish-and-chips shop in Exeter, in the green hill country of east Devon. Like all people who know fresh cod, she understood the importance of thickness. "If they don't let the small grow, that will be it," she added.

To most, though not all, of the British working class, *fish* means cod. In Yorkshire, though, it means haddock, and Harry Ramsden's, a fish-and-chips chain founded in Guiseley, Leeds, in 1928, built a reputation for more than sixty years with fried haddock. But in the 1990s, when Ramsden's expanded into the south of England, it was forced to switch to cod. Nothing else is acceptable in the south of England, except in London.

In Dickens's description of a thieves' den in back-street London, he includes the basic components of a working-class commercial district. The London fried fish trade began with the industrial revolution in the 1830s. Jewish merchants in the East End and Soho fried fish and distributed it from warehouses decades before "chipped" potatoes were added. The fish was, and still is, whatever was getting a good price at Billingsgate market—cod, haddock, plaice, hake, or even skate or dogfish. In recent years, dogfish has increasingly turned up in chip shops, but only in London will shop owners admit to this. In most of England, the disappearance of cod, and large cods at that, is a threat to a way of life.

Newlyn is a dark, brick, tumble-down-to-the-docks Cornish fishing port, a few miles from Land's End on the most seaward tip of England. There David Jewell, another fish-and-chip shop owner, said that he cannot always offer the real fish. Fish-and-chips, the common man's dish, must sell for a reasonably low price, and the high price of cod sometimes forces him to settle for pollock or whiting. Also, this most British of foods is often not British anymore. Fishy Moore's, one of the oldest chip shops in England, founded in Coventry in 1891, has given up on English cod. Coventry, in the Midlands, is as far inland as any place in England can be. Yet, for eighty years, the Moore family would go by train every morning to Skegness, on the other side of England, and buy freshly landed North Sea cod. In 1968, the family sold the shop. The current owner, Shaun Britton, who learned the business from his father, said, "It is almost impossible to get good British cod. What we see on docks has been on a boat for three days." Fishy Moore's buys frozen cod from Iceland, Norway, or occasionally the Faroe Islands.

If there is anything as basic and universal to the British working class as fried fish, it is xenophobia. So the proposition that foreigners may be depriving British workers of their cod is politically potent. To the British fishermen, and to many British people, that is exactly what the European Community, which is now the European Union, has done. This argument, of course, denies the long British history of overfishing and the fact that the dread Spanish supertrawlers, which are now so universally denounced, were a British invention. And the rights of fishermen to have free access to the sea, a principle the British fought for with such high-minded rhetoric in Icelandic waters, was somehow forgotten each time Brussels suggested a European partner should have rights in British waters.

According to the British government, 70 percent of the species in British waters are being overfished. In the North Sea, the catch dropped from 287,000 metric tons in 1981 to 86,000 in 1991. Like Canada's northern stock,

British cod are now reaching maturity at a much younger age than the normal three to five years, and large cod are increasingly rare.

In the 1990s, with North Sea fisheries in crisis, the action shifted to the Irish box. Soon the International Council for Exploration of the Seas (ICES), which for years had been warning about the dwindling stocks in the North Sea, began reporting a similar situation in the Irish Sea.

Within the European Union, fishing issues are settled by a much disliked bureaucracy of the Common Fishing Policy. In each country, each boat has a set quota on each species in each area of ocean each month. It is widely agreed that this system has failed to stop the decline of cod and other commercially valuable species. Politics and nationalism often play far greater roles than conservation in the decision-making process. For many years, scientists and European ministers agreed that hake was so menaced that the quotas had to be reduced by 40 percent. But at each annual meeting of European fishing ministers, the Spanish, for whom hake is a basic food, lobby to maintain the catch, and the reduction is never agreed on.

In a December 1994 meeting for European ministers, the Common Fishing Policy agreed to let forty Spanish fishing boats work the Irish box. After forty years of overfishing by their own trawlers, the fishermen of Cornwall and Devon had someone else to blame for their dwindling cod stocks. Even before the forty boats arrived, the word *Spanish* seemed to sit unkindly on everyone's lips in southwestern England. Maureen Whitehead at her Exeter fish-and-chips shop now knew whom to blame if her fish pieces seemed a bit thin. "There'll be no more chunky pieces if the Spanish take everything," she warned.

Far out in balmy, green Cornwall, with its lush vegetation warmed by the Gulf Stream, there was a growing obsession about the Spanish. European relationships are often mired in history, and the Spanish have never had a good name here. The Cornish recall without forgiveness that long before the two brutal World Wars against the Germans, and centuries of battles against the French, the narrow sloping streets of Newlyn had been sacked by marauding Spaniards who arrived in galleons. And now they were coming back.

The Spanish, with the largest fleet and little to offer in fishing grounds, are a favorite target in Atlantic fishing. It is seldom mentioned that they also have the largest market because of an unusually high per capita consumption. For centuries, Atlantic fishermen from the New England pilgrims to the newly independent Icelanders have been sustained by Spanish markets.

Few cod are landed anymore by the Spanish or even the Basque fishermen who began it all. The giant *bacalao* companies—which owned their own trawler fleets, landed salt cod from the banks, and dried and sold it—have all closed. Trueba y Pardo used to be a major company in Bilbao. In San Sebastián's port of Pasajes, there was PYSBE—Pesquerías y Secadores de Bacalao de España (Salt Cod Fishermen and Driers of Spain), founded in 1926. Both closed in the 1960s. In addition to fishermen and dockworkers, PYSBE had employed 500 workers in cleaning and drying alone. Trueba y Pardo had about 200 cleaners. These workers were almost all women, earning very low wages and no benefits, spending their days simply removing the dark gray membrane that had been the organ cavity lining. It was thought to be unattractive, and these workers cost the companies very little. Each woman could process 1,000 kilos (2,200 pounds) of fish in a day. Then the cod would be air-dried in the mountains. With modern salaries and benefits, companies could not afford such huge payrolls.

When the 200-mile limit was imposed, the Basques lost access to most cod grounds. By the time Spain entered the European Community in the 1980s, European waters had little cod. By 1990, only a few very old trawlers were rigged for cod fishing from Basque ports.

The banking and financial services that were established in Bilbao and San Sebastián because they had been trade centers, continued to flourish. In this commercial environment Basques made a transition from landing and processing cod to becoming importers.

A few miles up the Nerbioi River from Bilbao, in the town of Arrigorriaga, Adolfo Eguino's office overlooked the steep green Basque mountains. His windows were closed off by wrought-iron grilles fashioned in the shape of splayed salt cod. He was the director of Baisfa, one of the largest Basque *bacalao* companies. A small, tough-looking man in his fifties, he had a gruff manner that charmed more than it offended. Eguino grew up in Portugalete, an old seafaring town whose name means "port of Galleons." Some of the men who raided Newlyn may have been from there. As a boy, he didn't like school and at the age of fourteen dropped out to start his own business. He specialized in selling *bacalao* to small food stores, like the ones in the chain owned by his father. In this way he came to know the people at PYSBE and Trueba y Pardo, and from those companies he found six partners to start Baisfa. They now imported all of their salt cod from Iceland—1,500 tons a year. In 1994, the Icelanders stopped delivering to them by ship, instead sending the salt cod to Rotterdam, where it was put on a truck and driven down. This meant that rather than receiving a huge shipment every few months, one container arrived every week. "The secret of good salt cod is to not let it spend much time on the boats," said Eguino.

And yet, somehow, the people of Cornwall and much of England were convinced that the Spanish would take all their cod. Realistically, the Spanish were more interested in taking all their hake, but since cod was what the Englishmen cared about, it seemed to follow that cod was what the Spaniards would take.

Newlyn does not look like the Cornish towns on either side: Penzance and Mousehole. Those are resort towns where British vacationers practice that peculiarly British pastime of strolling the beaches and walkways, bundled in sweaters and mufflers. But Newlyn is a fishing town—or, increasingly, an out-of-work fishing town. By the

1990s, the long piers at the bottom of town where the trawlers tied up usually had one or two vessels being "decommissioned," cut up and sold for scrap. In an attempt to reduce the size of the British fleet, the government was paying fishermen to destroy their boats. But there was almost no work to be had in Cornwall except fishing. Mike Townsend, chief executive of the Cornish Fish Producers Organization, summed up the position of Cornish fishermen: "There is nothing else here. If they don't catch fish they will have no work, but if they keep catching enough to earn a living the fish will disappear."

William Hooper, fifty-five, the burly skipper of the 135-foot *Daisy Christiane*, said, "If they decommission this boat, I wouldn't have enough to buy a sweet." Hooper had been fishing out of Cornish ports for forty years. "The stocks are not what they were ten years ago," he said. "They are diminishing slowly all the time. All you can do to compensate is a bigger boat with a bigger net, more expenses, and you still can't catch what you did ten years ago."

Hooper first went to sea in 1955, when, he said, "the fish were knee-deep because of so little fishing during the war." He can no longer earn a living on a forty-foot boat like the one he had then. Now, as a share fisherman, he worked on a company-owned trawler, and, like all of his crew, he fished hard to earn a percentage of the sale of the catch. No individual fishermen could afford the cost of fuel and maintenance on a ship large enough to haul two five-ton nets, the size needed to catch enough fish to be profitable.

There was a growing movement among British fishermen to recognize the Common Fishing Policy as a failure and withdraw from it. Mike Townsend was one of the most outspoken leaders of this movement. "Sometime we have to say, 'Stop. We are not managing the stocks in a sustainable way.' " His argument was that the United Kingdom would be a better guardian of its own waters.

A debate raged over the fundamental tool of fishery management, the quotas, which were based on ICES attempts to monitor fishing populations. If groundfish were diminishing, fishermen were to fish less of them and at the same time increase their catch of the smaller fish on which they feed. Through quotas, man was attempting to artificially readjust the balance of the species while fishermen continued to earn a living.

But estimates of stock sizes were based on landings, the fish brought to market, and not on catch, the fish taken onto the boats, which was closer to the number of fish killed. As much as 40 percent of catches were being dumped back into the sea, even though most of these fish were already dead. Fishermen were radioed market prices to their boats at sea, and if the price dropped too low on a species, they would dump those fish overboard. Townsend, and many others, believed that the quotas bore no relationship to the actual state of the fish stocks. He laughed at questions of vanishing cod. "We have been plagued by cod. We don't know what to do with them." But fishermen, including Hooper, did not agree. According to Hooper, though there were momentary increases, the stocks have been declining.

In 1995, Canadian fisheries minister Brian Tobin offered a diversion from the frustrating complexity of fishery issues when he arrested a Spanish trawler, the *Estai,* confiscating the ship, the catch, and the gear. The *Estai* was held for a week while Canadian fisheries authorities rummaged through the 350 metric tons of fish aboard for evidence that its captain had violated North Atlantic conservation standards. Then, armed with photos of undersized Greenland halibut, also known as turbot, which the *Estai* had caught in the international part of the Grand Banks, and a salvaged undersized net, which the Spanish had dumped overboard, Tobin went to the UN in New York. There he delivered a defiant speech asserting that Canada intended to continue arresting Spanish trawlers and cutting their nets until some international conservation policy was established for the waters beyond Canada's 200-mile limit. He went on to say that Canada was taking this stand with humility, recognizing its own guilt for overfishing in the past. He said that Canadians took no pride in doing this, a rhetorical embellishment that the Canadian press seized on. The *Toronto Star* said, "Tobin was a bit disingenuous there. Canadians are proud—even gleeful—at the sight of one of their politicians finally standing up and doing something about one of the world's environmental disasters. In grim cost-cutting times, his colorful language and flare for the headlines have made Tobin the best act in town."

Tobin called the Spanish captains "rogue pirates," and Newfoundland premier Clyde Wells held up photos of undersized fish while accusing the Spanish fishermen of lying and cheating. The Canadian Coast Guard continued to chase Spanish trawlers off the international section of the Grand Banks, and when one trawler moved back in, Canada stole a page from Iceland's Coast Guard and cut its net. The Canadians were very happy. The British were very happy. In Newlyn, they flew the red maple leaf flag of Canada. Fishermen from Newfoundland to Rhode Island to Cornwall cheered. In the 1996 election, Tobin was voted Newfoundland premier in a big victory for the Liberals. The Europeans also had their victory. Emma Bonino, fisheries commissioner for the European Union, said Canada's fisheries minister was the real pirate and denounced Canada for reckless acts endangering the lives of Spanish fishermen on the high seas. The Canadians quietly released the men, ship, and gear, and the Spaniards, who had their own electorate to think of, threatened to sue Canada. Politically, the incident was a win for all sides.

In Petty Harbour, Sam Lee said, "It was good to watch, but it wasn't real. It was like going to the movies."

To the Cornish fishermen, it was a further vindication for their survival struggle against the Spanish. William Hooper said, "The biggest problem we have is the Spanish." He was asked how it could all be the fault of the Spanish since they were newcomers and the catch had been declining for forty years. Hooper thought a minute and then added, "Yes, the Scots used to overfish."

BACALAO—NATIONAL AND INTERNATIONAL

BACALAO A LO COMUNISTA (COMMUNIST-STYLE SALT COD)

Divide the salt cod into thin filets, then dredge each in flour and then fry them. In a baking dish put a layer of salt cod, then a layer of sliced potatoes and parmesan cheese. Cover with béchamel sauce and gratin in the oven.

—*El Bacalao,* the recipes of PYSBE
(Salt Cod Fishermen and Driers of Spain),
San Sebastián, 1936

BACALAO BANDERA ESPAÑOLA (SPANISH FLAG SALT COD)

Choose the best salt cod, boil it, remove the skin and bones, flake it, and put the flakes in a serving dish. Make a good mayonnaise with garlic. Put it over the salt cod, completely covering it, and the width of the dish. On either side place strips of red pepper, roasted or fried, thus resulting in the colors of the Spanish flag. Each pepper strip has to be half the width of the strip of mayonnaise.

—Alejandro Bon,
Leonor, Superior Cook, Barcelona, 1946

14: Bracing for the Canadian Armada

THE TIME MUST COME WHEN THIS COAST WILL BE A PLACE OF RESORT FOR THOSE NEW ENGLANDERS WHO REALLY WISH TO VISIT THE SEA-SIDE.

—Henry David Thoreau,
Cape Cod, 1851

Today, Gloucester has as much in common with its neighbor on Cape Ann, Rockport, as Newlyn has with Mousehole. Rockport is a pretty little town with a pretty little harbor full of expensive yachts. The waterfront shops sell crafts and snacks for "New Englanders who really wish to visit the sea-side." Gloucester could have been Newlyn's sister. It is a rough, downhill fishing town. Fine old wooden merchant's houses view the sea from up on the hills, while nineteenth- and twentieth-century brick buildings—the look of old blue-collar New England—dominate the lower part of town around a well-sheltered and busy waterfront. Bottom draggers, a few longliners, gillnetters, and lobster boats line the docks. In early morning they head out, a few at a time, and from four o'clock on they come back, trailed by gulls as they make their way with their catches toward the landing docks of the seafood companies. The companies are small. Birdseye's old company, which became Postum, which became General Foods, was then sold to O'Donnell Usen, which left for Florida. Seafood companies didn't need to be in fishing ports anymore. Their fish arrives in freezer containers, often from other oceans. Gorton's is still in Gloucester, the largest plant with the biggest sign, but the company hasn't bought a fish from a Gloucester fisherman in years. Gorton's buys no Atlantic cod from anyone anymore. In 1933, with the invention of the filleting machine, redfish, which had always been tossed overboard, became a major catch, and by 1951 represented 70 percent of all fish landed in Gloucester. But in 1966, Gorton's bought its last Gloucester redfish too, closing down the plant on what had been called "redfish wharf."

By the mid-1990s, the town had about 400 working fishermen left, down from 2,000 forty years before. Gloucester's fleet had the fatal flaw of being picturesque. There were too many old wooden hulled trawlers, which insurance companies won't even cover anymore, or rusting old steel ones and too many little low-built gillnetters. They gave a wonderful look to the old harbor, but it meant that the Gloucester fleet was not modern. But maybe not modernizing was the way of the future.

The New England Fishery Management Council was charged with the task of holding back the fleet from scooping up the last of the groundfish in New England waters. The Magnuson Fisheries Conservation and Management Act of 1976 had extended the exclusive U.S. fishing zone to 200 miles offshore and set up as regulators regional fishery management councils dominated by fishing interests. Fishermen never had been good regulators, but they were virtually encouraged not to be by loan guarantees and other financial incentives that led to a massive growth in the U.S. fishing fleet. In 1994, when the National Marine Fisheries Service counted fish stocks, it concluded that the fleet was about twice as large as the fish stocks could sustain. The assessment showed that the cod stock on Georges Bank was about 40 percent of what had been found in 1990. That sharp a decline had never before been measured on Georges Bank. "This really got the attention of the New England Fishery Commission, and that is how tougher measures got through," said Ralph Mayo of the National Marine Fisheries Service.

Each vessel was restricted to 139 days of groundfishing annually. The goal was to take only 15 percent of the stock in a year of fishing. But in 1996, when it was calculated that in those 139 days fishermen had taken 55 percent of the stock, restrictions were further tightened to 88 days. This system of conservation greatly favors small boats over the large trawlers. The owner of a large bottom-dragging trawler has enormous maintenance costs, such as $30,000 or more per year for insurance, and cannot afford to have his vessel sit idle 277 days each year. Most fishermen said that even in the winter, when the groundfishing was good, they would still rather crew on a small gillnetter than on a big trawler, because the catch wasn't enough to split among a six- or seven-man trawler crew. If fishery management could actually force out larger boats, it would greatly reduce the capacity of the fleet, and this could be part of a solution.

Georges Bank was the one bank that still had cod fishing. Canada had won rights to a part of what is called the Northeast Peak, which Canadians fished from June to December. After 1994, the United States closed its part of the Northeast Peak, but the western part of Georges Bank was still fished with some success. Since they were severely limited in the number of days they could go groundfishing, Gloucester fishermen began asking the government for financial help—the same kind they had gotten to build their bottom-dragging fleet after the 200-mile limit was established—to convert their vessels to midwater trawlers.

The seas seemed suddenly full of *pelagic* fish—midwater species such as herring, mackerel, and menhaden. Since these fish were normally eaten by the now-vanishing cod, the two phenomena might have been related. Ralph Mayo rejected this theory, pointing out that the herring boom began in the late 1980s, before the cod decline. Or at least before the cod decline was perceived.

Just as a codfish would do, the fishermen simply

turned toward the available food source. In the 1960s, skate was sold to lobstermen as bait for one dollar a bushel, herring was cheap bait for longlines, and dogfish were the curse of gillnetters. The rough skin of dogfish was hard on fishermen's hands and so difficult to get disentangled from the nets, fishermen would hack them out with knives and hose the gory mess overboard. By the 1990s, herring, skate, and dogfish were all target species.

In the 1990s, dogfish, marketed under its new name, cape shark, though still low-priced, was selling well, especially for export to Europe and Asia. In fact, by the mid-1990s, it no longer seemed likely that the dogfish would take over the cod's niche in the food chain, because these little sharks themselves were being somewhat overfished. A shark is not a fish, and instead of laying millions of eggs every year, a dogfish gives birth to five or six "pups" every other year. It is not biologically capable of withstanding the siege cod has faced.

Truck drivers, repairmen, dockworkers, and captains of tour boats—all over town there are ex-fishermen. All the men who work on the dock for Old Port Seafoods are former fishermen. Dave Molloy, a small fit Gloucester native in his forties, had grown up fishing with his father. In 1988, he gave up. "I knew it was over. I fished for seventeen years, but the last three years I starved."

The concrete pier of Old Port Seafoods has two unloading cranes, small rope-and-pulley affairs with a motorized drum to give a lift to the rope. Inside, a man and some women stand at a stainless steel counter, filleting

Schooners in Gloucester harbor, early twentieth century. (Peabody Essex Museum, Salem, Massachusetts)

small cod into scrod. The man wears a Red Sox baseball cap. Cod and the Red Sox—Massachusetts's beloved losers. At either end of the pier, rusting steel-hulled trawlers are tied up, their nets rolled up high off the stern, waiting for a better day.

Instead, small boats come in with the strange new catches. A gillnetter arrives with a small load of herring, quickly shoveled up, unloaded, weighed, and iced. The next boat in, Russell Sherman's seventy-foot trawler, has its story told by the birds that do not even bother to follow the vessel. Even they are not interested in a few flatfish and some monkfish, another species New Englanders recently learned to keep. "Just scraps, but at least my boat leaves the dock," Russell says. "I'm not going to use up my days with these fish. I'm fishing three-inch mesh. I'll wait until winter and then put on six-inch mesh and go after them [cod]. It's not worthwhile to change until winter."

Such sad tales turn fishermen's talk toward cod. Dave Molloy, operating the forklift, says, "You want to know about cod, I'll tell you." He puts up his hand and pretends to whisper. "There ain't no more."

"It's coming back," Russell insists. "They should have done this twenty years ago. We'd have cod out our assholes by now. We should have used six-inch mesh twenty years ago. Like Iceland did."

"I said that years ago before the magnimity imported" (translation: before the importance was realized), another fishermen asserts in that language which is found only along the southern New England coast. While Russell Sherman gets into a conversation with an older Sicilian fisherman about his struggle to lose weight, Nicki Avelas, another former fisherman, who is a part owner in the seafood company, tallies up the small catch. Nicki's big blond dog follows the action closely and finds good bits to eat. After Russell finishes unloading and gets his receipt, he hoses down his deck and shoves off.

"See you tomorrow," says Dave Molloy, tossing him his bowline. As the boat putters out of Gloucester harbor disappearing behind a row of idle bottom draggers, Dave shakes his head and says, "That guy hasn't made a dime all summer." It is September.

Waiting for the next boat, they grumble, as everyone

in Gloucester is doing this week, about a large Russian factory ship in the harbor. It is no longer allowed to fish U.S. waters, but it came in to buy from fishermen. For all their complaining, the fishermen always sell to them. One fisherman accuses the ship of flying the red flag higher than that of the United States, even while in port. He insists he has seen this and does not seem to know that Russia doesn't use a red flag anymore.

A fifty-foot three-man gillnetter comes in. The captain, Cecil, is almost as wide as he is tall, his blond hair lighter than his weatherbeaten face. One of the three crewmen is his son, a young man with the same build. They have been out following the tuna boats, which throw over chum. The dogfish chase the chum, and they, having set their nets overnight in the tuna grounds, are now coming in with their deck packed with bleeding little sharks. Dogfish pay only thirteen cents a pound, seventeen for the larger ones, and most of this catch is small. As Cecil works the ropes and a huge crate of the slimy catch swings over the pier, someone jokingly shouts, "There it is, fish-and-chips for London."

Gloucestermen claim that bluefin tuna had been running well and that all the talk about it being rare was a ploy by sports fishermen. "Environmentalists and sports fishermen. It's a highbrow thing." Enough tuna is getting landed so that it serves as the logo for Old Port Seafood. As long as fishermen can catch a fish, they resist the idea that the species is in trouble. But with cod, they all recognize that there is a problem. Except Nicki, who argues, "There's lots of cod out there. If you used a three-inch net, you would get lots of legal cod. A twenty-inch fish escapes a six-inch net. If they just kept the regulations in place, the fish would come back. If they keep adding restrictions, the fish will be back but the fishermen will be gone."

In Gloucester, it is a commonly held belief that the damage from overfishing is only temporary but that the restrictions are doing permanent damage to the community. Soon, it is believed, the cod will be back, and the fishermen will be gone, their boats turned into scrap. And then—and this is the maddeningly unjust part—according to this scenario, the Canadians, their historic competitors, are going to come down and take all their fish.

Bad blood between Canadian and New England fishermen dates back to the French and Indian War when French fishermen from Cape Breton had menaced New Englanders and Gloucestermen fought with the troops who took the garrisoned French fishing station at Louisbourg. Nova Scotians and Quebecers had refused to side with New England in the Revolution. In 1866, both the British and the Canadians had excluded New Englanders from the Canadian three-mile limit. In 1870, five Gloucester schooners had been seized by Canada, and the Gloucester citizenry had petitioned Congress to cut relations with Canada. The *Edward A. Horton*, the Gloucester schooner forced into Guysborough, Nova Scotia, and stripped of its sails, is part of Gloucester lore. Six New Englanders broke into a warehouse at night, took the impounded gear, rerigged the schooner, and slipped away on the flood tide.

Of course, during most of this history, the ancestors

of most of the present-day Gloucester fishermen were jigging off of Sicily, along the Greek islands, or off the Azores. But more recently, when Gorton's had closed its redfish operation in Gloucester, the company had moved it to Canada. And when the 200-mile limits were established, New Englanders had fought to keep Nova Scotians off Georges Bank, while Nova Scotians had fought to keep them off the Grand Banks. This fear of Canadian competition is part of Gloucester culture, the same way fear of Spaniards is somehow in the brick walls of Newlyn.

It is true that fishing policy is forcing fishermen out. Angela Sanfilippo, a leader in the activist group Fishermen's Wives of Gloucester, organized a program to retrain fishermen for other jobs. After two years, she has found new jobs for twenty-nine fishermen—as marina workers, truck drivers, mechanics, plus a few jobs in the computer field. But her own husband, John Sanfilippo, told her, "No one is ever going to stop me from fishing."

Like many Gloucester fishermen in the late twentieth century, the Sanfilippos are from Sicily, where catches were meager and boats small. John, born in 1945, the ninth child of a fishing family, began in a little dory with his father. They gillnetted, longlined, jigged, purse seined, and survived in the postwar years of poverty. He moved to the United States when he was twenty-two. Angela came in 1963 as a seventeen-year-old. The men in her family also had been fishermen for generations. She had relatives salmon fishing in Alaska, and tuna fishing out of San Diego. Her parents took her to Milwaukee, where cousins were fishing the Great Lakes. But the fish were dying from pollution, and the experience left Angela with a keen sense that polluters were the enemy of fishermen. Her father got a job in a foundry to support the family, but unable to give up fishing, he went out on weekends. Deeply unhappy, the family was about to return to Sicily when friends told them about Gloucester.

When John came to Gloucester, he abandoned all other forms of fishing for bottom dragging. Groundfish were the prize, though each Sunday, his day off, he fished for bass on the State Pier. Most fishermen cannot stop fishing. Lobstermen will take a rod and reel and try some trolling while waiting for tides. When fishermen of the Portuguese White Fleet had a full hold of cod, and put in at St. John's for supplies for the return journey, they would come ashore to catch trout in the streams. When Angela was pregnant with their daughter, John became restless and bought bait from a Russian factory ship to go longlining for swordfish with buoys made of chlorine bottles. On August 3, 1975, the night their daughter was born, he caught sixty-five swordfish in the deep water beyond 200 miles known as the Canyons. Once their children were grown, the Sanfilippos finally took a vacation to Bermuda. John went fishing with two poles off the pier of the Princess Hotel. Using french fries for bait, he caught tropical fish that resembled the rockfish in Sicily. He advised the kitchen on how to prepare them.

They managed to send their son, Dominic, to Tufts, where he was a political science major. But after two years, he returned to Gloucester saying he wanted to be a fisherman. Angela cried. In Newfoundland, Sam Lee fought with his son because he also dropped out of school and wanted to fish.

"But after a couple of months," says Angela, "I realized that he is happy. He said he wanted to go to Georges Bank. He couldn't go before, because it was too far for the small boat with his father. So he crewed on a big dragger and fished the Bank just before it closed. Now he wants to buy a fishing boat. I tell him to keep his money. He will need it for something else. He says, 'I miss my sunrise and my sunset and the seagulls flying over me.' "

Vito Calomo, a Sicilian-born ex-fisherman who now works for the Fisheries Commission in the Gloucester Community Development Department, says, "You buy out a man whose father and grandfather were fishermen, and you are wiping out a hundred years of knowledge. A fisherman is a special person. He is a captain, a navigator, an engineer, a cutter, a gutter, an expert net mender, a market speculator. And he's a tourist attraction. People want to come to a town where there are men with cigars in their mouth and boots on their feet mending nets. We are going to lose all that."

At that moment, a pickup truck with a lawn-mowing tractor on the back comes down the coastal road, and Calomo shouts at the driver. "That's my brother. He was a captain, and now he's cutting grass. A captain, cutting grass. I saw one washing dishes in a restaurant and one who works as a security guard."

To Calomo, Sanfilippo, and most of the people in the Gloucester fishing community, their plight is not their fault but the responsibility of government. "What do they do about the Red Sox?" argues Calomo about Boston's perennially losing baseball team. "They don't get rid of the Red Sox. They fire the managers."

Calomo says, "Canada is going to be American, and we are going to be Canada. Because they are subsidizing out-of-work fishermen, they will have them when the fish come back. They are keeping their fishermen. They are going to fill our market. Who's going to be left to fish here when the fish come back?"

Angela Sanfilippo, who was active in the fight to stop oil exploration on Georges Bank, says, "Who is going to look after the sea if the fishermen are gone?" It is not an unreasonable question. Will it be Unilever, the huge multinational that bought Gorton's? Will Unilever launch an angry protest when a corporation pollutes the sea?

Is it really all over? Are these last gatherers of food from the wild to be phased out? Is this the last of wild food? Is our last physical tie to untamed nature to become an obscure delicacy like the occasional pheasant? Is Gloucester to become a village of boutiques, labeled "an artist colony," like Rockport? Will Newlyn one day be only for strolling, like its neighboring towns, or as has already happened to St. Sebastián? Will Gloucester harbor, too, be converted into a yacht basin? Or should it be preserved, as is Lunenburg, Nova Scotia, as a museum to the days of fishing?

Governments understand that there is a social function to having fishermen and having fishing ports. Even while they have programs to reduce the size of their fleets in order to save fish stocks, they are also subsidizing fishing because there is no work available for most ex-fishermen. In the developed world, only Iceland expects fisheries to make a serious contribution to the economy, and even that country is trying to reduce the number of fishermen. A 1989 study by the United Nations Food and Agriculture Organization estimated that it cost about $92 billion to operate the global fishing fleet. Revenue, on the other hand, was only $70 billion; much of the difference was made up by subsidies from governments to fishermen and boat builders. According to the FAO, by the early 1990s, the twelve-nation European Union was spending about $580 million in annual fishing subsidies, while Norway alone paid out about $150 million. The Japanese government was estimated to have extended $19 billion in credit to its troubled fishing industry, much of which credit will never be paid back.

Miles from Gloucester harbor, at the hotels along the rocky New England coastline—rocks once valued for drying cod and now loved as a scenic element—tourists eat their breakfasts and plan their day. In the distance, lobster boats and small trawlers glide by, their diesel engines out of hearing range. Many of the tourists are planning to go "whale watching." They talk of whales as adorable pets, how they flop and dive and make a real snoring noise. On this rugged coastline where fortunes were once made hunting whales, whale watching has become a prosperous business during the tourism months. The skippers of the whale-watching boats are usually out-of-work fishermen.

There is a big difference between living in a society that hunts whales and living in one that views them. Nature is being reduced to precious demonstrations for entertainment and education, something far less natural than hunting. Are we headed for a world where nothing is left of nature but parks? Whales are mammals, and mammals do not lay a million eggs. We were forced to give up commercial hunting and to raise domestic mammals for meat, preserving the wild ones as best we could. It is harder to kill off fish than mammals. But after 1,000 years of hunting the Atlantic cod, we know that it can be done.

A Cook's Tale

ONE MIGHT SAY THAT IT [COD] IS THE ONLY FOOD, APART FROM BREAD, WHICH, ONCE ONE HAS GOT USED TO IT, ONE NEVER GETS BORED OF, WITHOUT WHICH ONE COULD NOT LIVE AND WHICH ONE COULD NEVER EXCHANGE FOR ANY DELICACY.

—Elena Ivanovna Molokhovets,
A Gift to Young Housewives, St. Petersburg, 1862

Six Centuries of Cod Recipes

THE CORRECT WAY TO FLUSH A COD

"YES, YES, I WILL DESALINATE YOU, YOU *GRANDE MORUE!*"

—Émile Zola, *Assommoir,* 1877

There is no general agreement on how to resuscitate stockfish or saltfish. No two pieces of cured cod are of the exact same thickness, dryness, or saltiness, and furthermore, different people prefer different tastes, often depending on the type of dish being made. Soaking will generally take more than 24 hours, but for very dry stockfish it can be several days. Most cooks agree that the only way to know when a cured fish is ready for cooking is to break off a piece and taste it. The more it has been dried, the longer it must be soaked. Salted fish needs to have the water in which it is soaking changed periodically so that the fish is not sitting in salt water.

Hannah Glasse in the 1758 edition of her British book wrote that stockfish should be soaked in milk and warm water. Most modern cooks insist on cold water and many believe it is best when soaked in a refrigerator, especially during warm weather. Others have been known to turn to another modern invention, the flush toilet.

Deep inland in France, *La France profonde,* as the French like to say, on the far side of the mountain range called the

Massif Central, is the Aveyron. It is a rugged region of high green sheep pastures, deep gorges, and jagged rock outcroppings, the most famous of which, in Roquefort-sur-Soulzon, provides the natural caves for aging the world's most famous cheese. An isolated area where shepherds still speak a local dialect, the region would get supplies all the way from distant Bordeaux on river barges. Barges would move up the Garonne to the Lot to Rodez and other towns in the Aveyron. The stockfish, bought in Bordeaux and dragged in the river behind the barge for the two-day voyage, would be soft and ready for cooking when it arrived.

In the twentieth century, the Lot became increasingly polluted and unnavigable, but a new invention was well suited to the preparation of stockfish: the flush toilet. In 1947, the president of the Conseil, the governing body of France, asked his valet to flush the toilet once an hour for the next week in preparation for a special dinner he was preparing on Sunday. The dish was stockfish. The toilet was fed by a water tank mounted high up on the wall, the *chasse d'eau.* A stockfish left in the *chasse d'eau* for two days was soft and ready for cooking. The system was also ideal for salted fish, since the water was easy to change. All of this may be deemed unaesthetic, but, unfortunately, it is now more hygienic than using the Garonne and its tributaries.

Two Views of Stockfish

[Stockfish is] hard as lumps of wood, but free of bad flavor, in fact, without much flavor at all . . . though very nice as an appetizer, and after all, anything that performs that function cannot be all that bad.

—Poggio Bracciolini (celebrated Latin scholar), 1436

DRIED FISH IS A STAPLE FOOD IN ICELAND. THIS SHOULD BE SHREDDED WITH THE FINGERS AND EATEN WITH BUTTER. IT VARIES IN TOUGHNESS. THE TOUGHER KIND TASTES LIKE TOE-NAILS, AND THE SOFTER KIND LIKE THE SKIN OFF THE SOLES OF ONE'S FEET.

—W. H. Auden and Louis MacNeice,
Letters from Iceland, 1967

BEAT IT

Before the toilet and the refrigerator, the tool that seems inevitably tied to stockfish was the hammer. If stockfish is of good quality, it resembles a rough-hewn, soft wood a bit lighter than balsa. The fibers have to somehow be broken down.

Item, when it [cod] is taken in the far seas and it is desired to keep it for 10 or 12 years, it is gutted and its head removed and it is dried in the air and sun and in no wise by a fire, or smoked; and when this is done it is called stockfish. And when it hath been kept a long time, and it is desired to eat it, it must be beaten with a wooden hammer for a full hour, then set it to soak in warm water for a full 12 hours or more, then cook and skim it very well like beef.

—Author unknown,
Le Mesnagier de Paris, circa 1393

KILL IT: LUTEFISK

Norwegians soften stockfish to almost jelly by putting it in lye.

First the beaten stockfish is put in cold water for four or five days, but the water must be changed regularly. Then lye or pure, crumbled ash made of nothing but birch or beech is boiled in water in a pot and then set aside until the ashes fall to the bottom: then cold water is poured out of the pot into another container, where it stands until it is very clear. The fish is put in this clear water where it stays for three days and taken out of it three hours before it is to be washed in cold water and boiled like any other fish and eaten with melted butter and mustard.

—Marta María Stephensen,
A Simple Cookery Pocket Booklet for Gentlewomen, 1800
(translated by Hallfredur Örn Eiríksson)

Diversionary Tactic

In 1982, British novelist Graham Greene, an elderly resident of Nice, started making seemingly paranoid public accusations about corruption in city hall. It was suggested that the famous author of intrigue was beginning to lose his grasp on reality. But when asked about Greene's allegations in an interview, the mayor, Jacques Médecin, son of another famous Nice mayor, began talking about cooking and offered a recipe for stockfish. In time, the mayor slipped away to South America, where excellent salt cod is available but little in the way of true stockfish.

The following recipe, according to Médecin, who is not always taken at his word, was given to his father by a local fisherman named Barba Chiquin, which in dialect means "uncle who likes a good bottle." Barba Chiquin would invite children over for this dish.

Take a dry stockfish, pound 100 grams on a stone with a hammer, reducing it to a kind of powder. For 100 grams of stockfish, crush 4 cloves of garlic in a mortar. Heat olive oil in a skillet until it smokes and brown 2 pébréta *[hot peppers]. When the oil starts to smoke, toss in the mixture of dried stockfish and garlic. When this preparation is lightly browned, spread it on a piece of* pain de compagne *[country-style bread] and wolf it down.*

—Jacques Médecin, ex-mayor of Nice

Médecin warned against trying the recipe unless you have a well-stocked wine cellar to deal "with a thirst which will last at least four or five days."

Also see page 61.

THE BAD NEWS AT WALDEN POND

IT IS RUMORED THAT IN THE FALL THE COWS HERE ARE SOMETIMES FED COD'S-HEAD! THE GODLIKE PART OF THE COD, WHICH, LIKE THE HUMAN HEAD, IS CURIOUSLY AND WONDERFULLY MADE, FORSOOTH HAS BUT LITTLE LESS BRAIN IN IT,—COMING TO SUCH AN END! TO BE CRAUNCHED BE COWS! I FELT MY OWN SKULL CRACK WITH SYMPATHY. WHAT IF THE HEADS OF MEN WERE TO BE CUT OFF TO FEED THE COWS OF A SUPERIOR ORDER OF BEINGS WHO INHABIT THE ISLANDS IN THE ETHER? AWAY GOES YOUR FINE BRAIN, THE HOUSE OF THOUGHT AND INSTINCT, TO SWELL THE CUD OF A RUMINANT ANIMAL!—HOWEVER, AN INHABITANT ASSURED ME THAT THEY DID NOT MAKE A

PRACTICE OF FEEDING COWS ON COD-HEADS; THE COWS MERELY WOULD EAT THEM SOMETIMES.

—Henry David Thoreau,
Cape Cod, 1851

Thoreau made these observations on a trip to Cape Cod the same year that Herman Melville's *Moby Dick* would describe Nantucket cows wandering with cod's heads on their feet. Thoreau was right that the heads were not likely to be offered to a cow, but the reason was that people like to eat them.

NOT THE LIPS: FRIED COD HEAD

Obtain 4 medium size cod heads. More for a large family. After they have been sculped—(to sculp heads: with sharp knife cut head down through to the eyes, grip back of head firmly and pull)—prepare to cook as follows:

Cut heads in two, skin and remove lips. Wash well and dry. Dip both sides of head in flour, sprinkle with salt and pepper to taste. Fry in fat until Golden Brown on both sides. Serve with potatoes and green peas, or any other vegetable preferred.

—Mrs. Lloyd G. Hann, Wesleyville, Newfoundland,
from *Fat-back & Molasses: A Collection of Favourite Old Recipes from Newfoundland & Labrador,*
edited by Ivan F. Jesperson, St. John's, 1974

NOT THE EYES: FISHERMAN'S COD-HEAD CHOWDER

8 cod heads, the eyes removed
3 oz salt pork
2 sliced onions
6 sliced potatoes
butter
salt and pepper

Try out (render) the pork. Add the onions and fry until golden. Lay in a kettle, then add the cod heads and potatoes. Cover with cold water and cook till the potatoes are done. Season; add a good chunk of butter.

Fishermen think removing the bones is sissy. Cod head of course contains the cods' tongues and cheeks. Sometimes, too, the cod's air sacs, known as the "lights" or "sounds," were fried in salt pork and then added to the chowder.

—compiled by Harriet Adams, comments by N. M. Halper,
Vittles for the Captain: Cape Cod Sea-Food Recipes,
Provincetown, 1941

Also see pages 46–47.

Marblehead

In 1750, Captain Francis Goelet claimed that Marblehead, Massachusetts, was famous for its large, well-fed children who, he said, were "the biggest in North America." According to Goelet, "the chief cause is attributed to their feeding on cod's head which is their principal diet."

Cape Cod kids don't use no sleds,
Haul away, Heave away,
They slide down hills on codfish heads.

—Sea shanty

Icelandic Wisdom

Until this century, the dried heads were carried inland by pony, with racks mounting sixty heads sticking out of either side. Both Norwegians and Icelanders pick them apart for snacks. "You know," said Reykjavík chef Úlfar Eysteinsson, "you just sit around the table talking and crr-r-ack"—he made a motion as though pulling apart a cod head.

In 1914, the entire practice of eating cod head was denounced by the prominent Icelandic banker Tryggvi Gunnarsson for that greatest of Nordic sins, impracticality. He said the food value was not worth the cost of production, and he demonstrated this in a mathematical formula that even calculated eating time. The director of the National Library responded with a treatise on the social values of eating cod head. Among other virtues, he claimed it taught forbearance, and he repeated the old Icelandic belief that eating animal heads increases intelligence. (Icelanders also eat sheep heads.)

Younger generations in Iceland don't eat dried cod head or sheep head very much, and there has not been a verifiable decline in intelligence.

THE ISLAND HEAD

In much of the salt cod–eating world, there are myths about cod heads because the head is rarely seen. According to a medieval Catalan legend, the cod's head is removed to conceal the fact that it is human. Though salt cod is a regular part of the Caribbean diet, few Caribbeans have ever seen a cod head. Carmelite Martial, a popular Creole cook in Guadeloupe who was born in 1919, said she never saw one. But her grandmother, who was born in 1871, had told Carmelite that she had a cod head locked away in a strongbox. What is more, the head had hair on it. "I never saw it," said Martial, who does not include cod head in her extensive cod repertoire.

SPARE PARTS

TWO WIZENED LITTLE BOYS, LOOKING MORE LIKE TINY OLD MEN, APPEARED WITH TIN CANS. THEY HAD ON KNEE BOOTS AND WADED AROUND ON THE EDGE OF THE WATER AMONG THE FISH HEADS. EACH HAD A POCKET KNIFE AND WHEN HE FOUND A HEAD OF HIS LIKING HE CUT A THREE CORNERED SLIT UNDER THE JAW AND TOOK OUT THE TONGUE. WHEN HE HAD HIS PAIL FULL HE DUMPED THE TONGUES OUT ON AN EMPTY TABLE AND FILLED HIS CAN WITH CLEAN SEA WATER. THEN HE WASHED THE PILE AND REPLACED THE TONGUES IN THE CAN. . . .

. . . WE WATCHED THE CLEANING OF THE FISH EVERY DAY FOR A WEEK AND NEVER FOUND OUR TASTE FOR COD ON THE DINNER MENU AT ALL IMPAIRED, BUT WHENEVER THE CARD SAID "BROILED CODS' TONGUES," TWO THIN, WAN FACES AND LITTLE BODIES STOOPED OVER A PILE OF CODS' HEADS APPEARED BEFORE US AND WE KNEW THAT WE COULD NOT POSSIBLY ORDER TONGUES.

—Doris Montgomery,
The Gaspé Coast in Focus, 1940

Once the meat, the head, and the liver have been eaten, is the rest ready to be ground into fish meal? From the rickety fishing villages of the Newfoundland and Labrador coasts, to rugged New England communities, to the fishermen's families of Brittany and Normandy, to the Basque women who washed salt cod for slave wages, to the pre-twentieth-century Icelanders who had almost nothing, come the following recipes, most of them delicacies today, although they originated with the poor.

Tongues and Cheeks

The scallop-sized, or sometimes even larger, disk of flesh on each side of the head is the most delicate meat on the cod. It is often served with "tongues," the throats, which have a richer taste and more gelatinous texture.

Cod tongues and cheeks, [are] rolled in corn meal, fried until brown. The tongue is not really the tongue, but the blob of meat at its base.

Pork chop is a cheek cut off with a piece of jaw bone and fried.

—compiled by Harriet Adams, comments by N. M. Halper,
Vittles for the Captain: Cape Cod Sea-Food Recipes,
Provincetown, 1941

STEWED CODFISH TONGUES

1 lb. fresh codfish tongues
1 large onion
½ lb. clear pork fat
salt
pepper

Place pork in fry pan and let cook until brown, add onion then tongues which have been cleaned well. Add salt and pepper to taste. Simmer about ½ hour.

—compiled by the Ingonish Women's Hospital Auxiliary,
From the Highlands and the Sea,
Ingonish, Cape Breton, Nova Scotia, 1974

The Basque Tongue: Kokotchas de Bacalao Verde

Basques are passionate about fish tongues, both salt cod and fresh hake. They use the Basque word *kokotxas* (pronounced cocoachas, it was sometimes spelled phonetically before modern Basque was established) and commonly prepare dozens of recipes. This is the best-known classic. The oil would always be olive oil.

> *Ingredients: 100 grams of salt cod tongues, garlic according to taste, parsley (the more the better), small onion, oil, milk*
>
> *Preparation: Soak the tongues for 24 hours, changing the water three times a day. Then pour out the water and drain them. Put in a casserole with oil, garlic according to taste, parsley and a little onion. Let them brown a little and then add the tongues. Give them a turn and turn off the heat and leave it ten minutes. Then put it back on a very slow heat and add three spoonfuls of milk. From time to time lightly stir the casserole and when you see that it is done, remove from the heat, and it is ready to be served.*
>
> —*El Bacalao,* the recipes of PYSBE (Salt Cod Fishermen and Driers of Spain), San Sebastián, 1936

Cod Roe

Fed to Frenchmen or to Fish

> *Roes of Cod well salted and Pickled are here neglected but are said to yield a good price in France to make Sawce withall.*
>
> *When the same are to be used, bruise them betwixt*

two trenchers, and beat them up with vinegar, White Wine etc. then let them stew or simmer over a gentle fire, with Anchovies and other Ingredients used for Sawce, puting the Butter well beat up thereto: We our selves on the Coasts use the Roes of Fresh Cod for Sawce.

—John Collins,
Salt and Fishery, 1682

Or to British Seamen . . .

Boil as directed (to every gallon of water add 1 gill of vinegar and 2 oz salt. Bring to a boil, put in the [roe], draw to the side of the fire and simmer gently till the fish is cooked.) and serve with parsley or caper sauce, or coat with egg or batter and breadcrumbs, and fry. Serve with quarters of lemon or anchovy sauce.

—C. H. Atkinson,
The Nautical Cookery Book for the Use of Stewards & Cooks of Cargo Vessels,
Glasgow, 1941

For Lent in Greece: Taramosaláta

Throughout the Christian Mediterranean, salt cod has remained a Lenten tradition. In Greece, *Taramosaláta* is served during lent. Since roe must be quickly eaten unless salted or smoked, it is generally a delicacy of northern nations. *Taramosaláta* was originally made from the roe of the golden gray mullet, which is native to the Mediterranean. But as Mediterranean fisheries declined, the Greeks started importing Norwegian cured cod roe, which they called *taramá.*

150 grams taramá *(salted cod roe)*
1 medium onion, grated or finely chopped
*1 slice (5–6 cm thick) stale bread**

*1 boiled potato**
juice of 1–2 lemons
1 cup olive oil

Remove crusts from the bread, soak it and squeeze dry. Rinse taramá *in water in a fine-meshed strainer to remove some of the salt.*

Pound the onion to a pulp in a mortar (goudi) *if you have one, then the bread and potato, then the oil and lemon juice, alternately, pounding or beating till smooth, or put it all in an electric blender. Spoon a little oil over the surface and garnish with olives.*

Some roe has a richer colour than others; one can cheat and add a little beetroot juice to improve the pale variety which in fact is of a finer quality.

*Only bread or only potatoes can be used, but the combination makes a good texture. Whole wheat bread gives a better flavour.

—Anne Yannoulis,
Greek Calendar Cookbook, Athens, 1988

Cod-sounds

To Broil Cod-Sounds

Clean and scald them with very hot water, and rub them with salt. Take off the sloughy coat, parboil them, then flour and broil til done. Dish them, and pour a sauce made of browned gravy, pepper, cayenne, salt, a little butter kneaded in browned flour, a tea-spoonful of made mustard, and one of soy. Cod-sounds *are dressed as ragout, by boiling as above (boil slowly in plenty of water, with a handful of salt) and stewing in clear gravy, adding a little cream and butter kneaded in flour, with a seasoning of lemon-peel, nutmeg and mace. Cut them in fillets. They may be fried.*

—Margaret Dods,
Cook and Housewife's Manual, London, 1829

Also see page 190.

Tripe

Stomachs are used as sausage casing. In Iceland, according to Hallfredur Örn Eiríksson, a folk customs scholar at the Árni Magnússon Institute, they are "cleaned thoroughly, stuffed with liver, which was sometimes kneaded with rye and then boiled and eaten. The same was sometimes done with sounds."

The tripe, the stomach lining, is also sometimes used. In 1571, a reception was given in Paris for Elizabeth of Austria. On the menu was Cod Tripe.

Hake and Squid with Cod Tripe Catalan Style

Ingredients for four servings: 4 choice center cut pieces of hake, 2 squids of 30 grams each after cleaning, 100 grams cod tripes, beef stock, 100 grams spinach, 1 spoonful of Corinth raisins, 1 spoonful of pine nuts, white beans, a sprig of chervil, 1 shallot, butter, virgin olive oil, red wine, salt and pepper.

Soak the beans, the raisins and the tripe the night before, each one separately.

Scale the hake but keep the skin on and wash it.
Clean the squid and remove the tentacles.
Cook the beans in water over a low heat.
Poach the tripe, saving the cooking water.
Cut the squid into julienne strips; saute in olive oil.
Put a pan on medium heat with butter and cook

the minced shallot. When it is soft, add red wine that has already been reduced. Then add the beans and the tripe, diced, then a little beef stock, and bring to a boil. Bind it with butter, salt and pepper.

Season the hake and put it in a pan with the skin side oiled.

Quickly saute the spinach, previously washed, and add pine nuts and raisins.

Make a bed of beans on each plate, add a spoonful of tripe, some sauce and the squid. On top of this, place the hake, skin side up. Decorate the edges with beans, spinach, pine nuts and raisins. Place a sprig of chervil on top of the fish.

—El Raco de con Fabes restaurant,
Barcelona, from Rafael García Santos,
El Bacalao en la cocina Vasca y las mejores recetas del mundo
(Salt Cod in Basque cooking and the
best recipes of the world), 1996

Down to Skin and Bones

Before Iceland was modernized, cod skin was roasted and served to children with butter. Hallfredur Eiriksson recalled from his childhood: "The skin is always pulled off the dried fish before it is eaten; the dry skin is tough but becomes soft and edible when roasted over the open fire."

Cod bones (as well as sheep and cattle bones) were prepared as follows:

> *[The bones] are put in sour whey where they lie until they are partly disintegrated and soft and then the whole thing is boiled slowly until the bones are tender and the mixture curds like thick porridge.*
>
> —Andrea Nikólína Jónsdóttir, Ný matreidslubók, 1858

CHOWDER

The word comes from the French *chaudière,* which was a large iron pot. Today the pots are often aluminum but are still standard equipment on fishing vessels, used for a simple warm, one-pot dish of fresh fish and ship's provisions. Most North Atlantic fishing communities make some variety of chowder. A sixteenth-century recipe for chowder was written in the Celtic language of Cornwall. The Cornish word for fishmonger, *jowter,* leads some historians to argue that chowder is of Cornish origin. It is frequently said that the French and English fishermen on the Grand Banks introduced chowder to Newfoundland cuisine and that from there it traveled south to Nova Scotia and New England. But Native Americans in these regions were already making fish chowder, though without the pork, when the Europeans arrived.

The original ingredients were salt pork, sea biscuit, and either fresh or salt cod, all carefully layered in the pot. These ingredients are standard long-conservation provisions of a fishing ship. Sea biscuits or hardtack, which later developed specific names for different shapes and sizes, such as pilot bread or Cross Crackers, were the forerunner of the cracker—a bread too hard to go stale. Potatoes were added to chowder recipes later. Newfoundland's Fishermen's Brewis is a classic chowder, but with the liquid cooked away. (See page 11.)

But Please . . .

NOWADAYS, ALL TOO FREQUENTLY IT [CHOWDER] COMES TO THE TABLE IN A THIMBLE. YOU MEASURE IT OUT WITH AN EYEDROPPER. YET, IN ITS DAY, A CHOWDER WAS THE CHIEF DISH AT A MEAL. THOUGH IT HAS FALLEN FROM THIS PROUD ESTATE, IT IS NOT, NOT, ONE OF THOSE FINE, THIN FUGITIVE SOUPS THAT YOU DELICATELY TOY WITH IN A GENTEEL LADY'S TEA-ROOM. . . . AND P.S.—PLEASE DON'T SERVE IT IN A CUP.

—compiled by Harriet Adams, comments by N. M. Halper,
Vittles for the Captain: Cape Cod Sea-Food Recipes,
Provincetown, 1941

Add What You Like

Four pounds of fish are enough to make a chowder for four or five people; half a dozen slices of salt pork in the bottom of the pot; hang it high, so that the pork may not burn; take it out when done very high brown; put in a layer of fish, cut in lengthwise slices, then a layer formed of crackers, small or sliced onions, and potatoes sliced as thin as a four pence, mixed with pieces of pork you have fried; then a layer of fish again, and so on. Six crackers are enough. Strew a little salt and pepper over each layer; over the whole pour a bowl-full of flour and water, enough to come up even with the surface of what you have in the pot. A sliced lemon adds to the flavor. A cup of tomato catsup is very excellent. Some people put in beer. A few clams are a pleasant addition. It should be covered so as not to let a particle of steam escape, if possible. Do not open it, except when nearly done, to taste if it be well seasoned.

—Lydia Maria Child,
The American Frugal Housewife,
Boston, 1829

In nineteenth-century New England, chowder parties became fashionable. A dozen or more people would go for a morning sail and then prepare the chowder either on board or later, on the beach. Also at this time, adding milk to chowder came into fashion, which meant that a chowder then required more than basic seagoing provisions. (See page 76.)

Fish Chowder: A New Approach

Fannie Merritt Farmer, an enormously influential cookbook writer, believed in extremely precise instructions and popularized the idea of exact measurements for recipes, an illusion of science that has become standard practice and, for more than 100 years, has left household cooks saying, "What went wrong? I followed the recipe." She was the most famous director of the Boston Cooking School, founded a generation earlier to teach working-class women how to cook "scientifically." Influenced by this school, freedom of choice has slowly been exorcised from recipes, and experimenting is increasingly discouraged.

Fannie Farmer's chowder recipe differs greatly from previous ones not only in its precise measurements but in that it is not made in one pot and completely abandons the idea of building a chowder in layers. The following recipe is clearly designed with a stove in mind, using several pots and even more than one burner at a time. When stoves replaced hearths, the way people cooked changed.

4 lb. cod or haddock.
6 cups potatoes cut in 1/4 inch slices, or
4 cups potatoes cut in 3/4 inch cubes.
1 sliced onion.
1 1/2 inch cube fat salt pork.
1 tablespoon salt.

1/8 teaspoon pepper.
3 tablespoons butter.
4 cups scalded milk.
8 common crackers.

Order the fish skinned, but head and tails left on. Cut off head and tails and remove fish from backbone. Cut fish in two-inch pieces and set aside. Put head, tail and backbone broken in pieces, in stewpan; add two cups cold water and bring slowly to a boiling point; cook twenty minutes. Cut salt pork in small pieces and try out, add onion, and fry five minutes; strain fat into stewpan. Parboil potatoes five minutes in boiling water to cover; drain, and add potatoes to fat; then add two cups boiling water and cook five minutes. Add liquor drained from bones and fish; cover, and simmer ten minutes. Add milk, salt, pepper, butter and crackers, split and soaked in enough cold milk to moisten. Pilot bread is sometimes used in place of common crackers.

—Fannie Merritt Farmer,
The Boston Cooking-School Cook Book, 1896

The Last Terre-Neuvas

In the French Channel ports, the men who fished Newfoundland were called the Terre-Neuvas. The last of them left from the Breton port of St.-Malo and the Norman port of Fécamp in the 1970s. In the campaign of 1961, the year after the following book was published, 22,000 tons of Grand Banks salt cod were still landed in Fécamp. The original French word for chowder, *la chaudrée,* has vanished, and here, the Marseillaise word is used for the soup.

Fécamp Bouillabaisse

Preparation: 30 minutes

Ingredients: 500 gr. of salt cod, 750 gr. of potatoes, 100 gr. of onion, a few branches of celery, 1 white of a leek, 2 cloves of garlic, 2 table spoons of tomato paste, 3 table spoons of oil, 1 bouquet garni, salt, pepper, chopped parsley.

Desalinate, poach and drain the salt cod, heat the oil in a pot. Toss in the chopped onion, the leek and the celery, also chopped. Let it cook for ten minutes. Peel the potatoes, cut them in thick rounds and cook them in the above preparation. When the potatoes are almost cooked, add the salt cod. Let it simmer slowly for ten minutes. Serve very hot, sprinkle with chopped parsley. Optional: add a little crème fraîche at the time of serving.

This recipe won the prix Terre-Neuve.

—Committee for Study and Information for the Development of Salt Cod Consumption, *Salt Cod: The Fish, Its Preparation, Its Nutritional, Culinary, and Economic Qualities*, Paris, 1960

THE DIASPORA OF THE WEST INDIA CURE

WEST AFRICA: STOCKFISH AND EGUZI

The slave trade left West Africa with a taste for cured cod, though to most West Africans, all that remains is a tradition of salting and drying local fish. Some West African towns, such as Kaolack, Senegal, offer a sight that has vanished from Gloucester and Petty Harbour—a shore covered with miles of fish flakes. Kaolack is inland but near the coast on the Saloum River and serves as a jumping-off spot for the headwaters of the Niger, a major artery of regional trade which moves this saltfish through the sub-Sahara and Sahara. But Nigeria has hard currency from oil and can import cod. Nigerians, especially Ibos, love dried cod, which they too call stockfish. This recipe comes from an Ibo who was born in the town of Bende near the Delta, and who now lives in the United States.

> *Wash the stockfish in hot water and soak it five minutes. Then boil it for several hours until it is soft. Then add goat meat. When the goat meat is cooked add* eguzi *[seeds of the green squash known in Nigeria as melon]. Add onions and minced* ukazi *[an herbal leaf]. Add crayfish. Then stir in* ugbo *[a thickener made from ground seeds, which have been cooked for hours until soft].*
>
> —Joy Okori, Washington, D.C., 1997

Brazil: Bacalhua com Leite de Coco

1 pound saltcod
1 freshly-grated coconut
4 tablespoons butter or oil
2 chopped onions
2 chopped tomatoes
2 or 3 drops hot pepper sauce
1 tablespoon dendê oil [a palm oil from Bahia]

Desalinate salt cod. Remove thick milk from coconut and reserve. To the residue add 2 cups hot water and remove thinned milk by pressing through a sieve. Fry saltcod in butter or oil with the onions and tomatoes and wet with the thin milk of the coconut. Cook over a low flame, occasionally stirring. When ready to serve, shake the pepper sauce on the fish, add the dendê oil and the thick coconut milk.

—Rosa Maria, *A Arte de Comer Bem, Rio de Janeiro, 1985*

Jamaica: Codfish Run Down

Today, "Run Down" is usually prepared with dark, oily, local fish. But the old-fashioned way was with salt cod. Alphanso McLean makes it for friends, though it is considered "too country" to be served at his place of business.

Grate coconut and let it sit in water. Force it through a strainer. Boil the strained liquid and keep stirring until oil comes to top. Add saltfish, onions, tomato and serve with yellow yam and green banana.

—Alphanso McLean, chef, Terra Nova Hotel, Kingston, 1996

Jamaica: Ackee and Saltfish

Elena rashly asked Violet to give us a native dish. She produced what is called "saltfish and ackee"— which I afterwards found described as a dish highly esteemed by the natives but less by other people.

—Edmund Wilson, *The Sixties,* 1993

It seems that much of Wilson's grumpiness on the subject stems from the fact that he got ackee poisoning. "I don't remember ever suffering in such a peculiar way as this," he wrote. Ackee is a West African fruit brought to Jamaica in 1793 by the infamous Captain Bligh, for whom it is named in botany—*Blighia sapida.* Like its namesake, ackee requires careful handling. The fruit, which hangs flame red from trees in the mountainous Jamaican countryside, must be fully ripe—that is, bursting open—to be safe.

Ackee and Saltfish is regarded by Jamaicans as their national dish, but the saltfish is now so expensive that Jamaicans joke that it is their "international dish"—only the tourists can afford it. Terra Nova Hotel chef Alphanso McLean serves Jamaican breakfast (Ackee and Saltfish with fried biscuits) on the wide and breezy hotel veranda, not so much to tourists, who seldom go to Kingston, as to affluent Jamaican businessmen and politicians. The fried biscuits are called johnnycakes and are the same biscuits served for breakfast with Jamaican molasses in the other Terranova, Newfoundland. Originally from southeastern New England, they were made from cornmeal and molasses, baked with pork dripping, and called jonny cakes, the name coming from "journey cakes" because they were taken on the road. They have followed the molasses-and-salt-cod route.

Caribbean saltfish dishes always involve shredding the fish, because it is of low quality. The saltfish, barely soaked, is hard and salty. The dishes depend on this for flavor.

> *Soak 1/4 pound salt cod for 20 minutes. Boil it for 10 minutes. Boil fruit from 1 dozen fresh ackee 5 minutes. Heat vegetable oil in a skillet. In the countryside we always used coconut oil, but here I use soy. Add chopped onions, scallion, thyme, and ground black pepper. That ground pepper gives it a nice flavor. Then add minced pepper [hot scotch bonnet pepper]. Add ackee and crumbled saltfish.*
>
> —Alphanso McLean, Terra Nova Hotel, Kingston

PUERTO RICO: SERENATA DE BACALAO

In San Juan, Puerto Rico, La Casita Blanca is just that—a one-story white building, a neighborhood bar built in 1922, which Jesus Perez took over in 1985. It is in Barrio Obrero, a neighborhood that many people do not want to go to after dark. But with its low one- and two-story houses in turquoise and salmon it is also one of the old areas of San Juan not yet overtaken by high-rise architecture.

Perez remembers that his family always made *bacalao* with roots, yams, breadfruit, yucca. "They ate it like this much more than with rice. My mother always bought whole fish hard and flat. Now I buy fillets. They are soft. They're salted but not dried." Drying makes the product more expensive, and since refrigeration is now widespread on the island, Puerto Ricans, and many other people throughout the developed world, cut costs by buying green cod. One salt cod dish from his childhood that has remained popular is *Serenata.* In St. Lucia this is called *Brule Jol;* in Trinidad *Buljol;* in Haiti, Guadeloupe, and Martinique *Chiquetaille.*

2 cups salt cod, desalinated, cleaned,
shredded, and boiled
1 large onion, sliced
1 garlic clove, minced
2 hot green peppers
½ cup stuffed olives
4 hardboiled eggs, sliced
2 boiled potatoes, peeled and diced
1 cup olive oil

Mix it well and serve with salt and pepper to taste.

—Jesus Perez, La Casita Blanca, San Juan, 1996

Also see page 91.

Guadeloupe: Feroce

Carmelite Martial, when asked what her favorite saltfish dish was, replied, "Well, since I don't really like saltfish, maybe a little *feroce.* I like avocados."

Mix avocado, kassav (cassava flour), grilled salt cod, a little hot pepper, and sunflower seed oil. Work them together with a spatula. Some people add cucumber, but it is not essential.

—Carmelite Martial, Le Table Creole, St. Felix, Guadeloupe

THE GREAT FRENCH DISGUISE

Te conozco, bacalao
Aunque vengas desfrazao
(I would know you, salt cod
Even if you were wearing a disguise)

—Cuban proverb

Since at least the time of Taillevent, salt cod has always been embellished with richness because it is harsh. Butter, olive oil, cream have been used—Icelanders pour the rendered fat of lamb kidneys over it. In 1654, Louis de Béchamel, marquis de Nointel, financier in the court of Louis XIV, having invested huge sums in the Newfoundland fishery, and finding the market weak in France because Frenchmen did not like this dried and salted old fish, invented a sauce for it, which is now called béchamel sauce. The sauce enjoyed tremendous popularity with salt cod and many other dishes. Originally it was a simple cream sauce with spices such as nutmeg. Later it was enriched with eggs:

Saltfish with Cream

Take good barrel-cod, and boil it; then take it all into flakes, and put it in a sauce-pan with cream, and season it with a little pepper; put in a handful of parsley scalded, and minced, and stove it gently till tender, and then shake it together with some thick butter and the yolks of two or three eggs, and dish it; and garnish with poached eggs and lemon sliced.

—Charles Carter,
The Compleat Practical Cook, London, 1730

Still later, flour was added. The sauce reached its height of complexity in the early twentieth century with Auguste Escoffier's elaborate 1921 recipe, which included chunks of veal. But a simpler flour-and-cream béchamel has remained a standard salt cod sauce in Portugal, Spain, Italy, New England (creamed codfish)—wherever salt cod is eaten.

BALLS

There is no single dish more common to all cod-eating cultures than the codfish ball. At the end of the nineteenth century, while the U.S. Senate debated a proposed pure food act, Senator George Frisbie Hoare, occupying the same august seat from which Daniel Webster had once extolled the virtues of chowder, rose and delivered a lengthy oration on "the exquisite flavor of the codfish, salted, made into balls, and eaten on a Sunday morning."

New England: Better Start on Saturday

Salt fish mashed with potatoes, with good butter or pork scraps to moisten it, is nicer the second day than it was the first. The fish should be minced very fine while it is warm. After it has gotten cold and dry it is difficult to do it nicely. There is no way of preparing salt fish for breakfast, so nice as to roll it up in little balls, after it is mixed with mashed potatoes dip it into an egg and fry it golden brown.

—Lydia Maria Child,
The American Frugal Housewife, Boston, 1829

FRANCE: MORUE EN CROQUETTES

The book in which this recipe appears was a ubiquitous classic in early-twentieth-century French households.

When your salt cod is cooked, as directed above (put the salt cod in cold water and cook. Remove from heat the moment it is about to boil, skim it and cover), remove the skin and the bones and prepare a béchamel sauce, which you mix with the salt cod, then let it chill; it must be cold enough so that your salt cod can be rolled into balls; to do that the sauce must be thick.

Prepare a dozen balls and roll them in fine bread crumbs, then dip them in beaten eggs, bread them a second time and put them in a very hot fryer. When they are a handsome color, remove them, stack them in a pyramid and sprinkle them with chopped parsley.

—Tante Marie, *La Véritable Cuisine de Famille*, Paris, 1925

ITALY: SALTED COD CROQUETTES

The Italian Tante Marie was Ada Boni, editor of Italy's leading women's magazine, *Preziosa*. Her cookbook first came out in 1928. This recipe is from the fifteenth edition, translated by Mathilde La Rosa.

1½ pounds soaked baccalà
3 anchovy filets, chopped
1 tablespoon chopped parsley

½ tablespoon pepper
1 tablespoon grated parmesan cheese
2 slices white bread, soaked in water and squeezed dry
2 eggs lightly beaten
½ cup flour
1 egg, lightly beaten

Boil fish in water 30 minutes and cool. Bone skin and chop fine. Add anchovies, parsley, pepper, cheese, bread and eggs and mix very well. Shape into croquettes, roll in flour, dip into egg, roll in bread crumbs and fry in olive oil until brown all over. Frying time will be about four minutes on each side. Serves 4.

—Ada Boni, *Talismano della Felicità,* 1950

Portugal: Sonhos de Bacalhau

1 cup shredded salted codfish
1 cup flour
1 cup water
1 tablespoon butter
salt and pepper to taste
3 eggs

Soak two pieces of salt dry codfish overnight. Save water. Shred fish with your fingers in very fine pieces. Measure water that you saved and bring to boil with fish, add butter and pepper, pour flour in and stir quickly until dough pulls from the side of the pan. Remove from heat and cool. Add eggs, one at a time, mix well. Fry in a deep skillet with plenty of hot oil, by dropping small spoonfuls in. Fry until golden brown. Makes about 20 to 24.

—Deolinda Maria Avila,
Foods of the Azores Islands, 1977

Jamaica: Stomp and Go

Mix 1 pound flour with water until it is thin.
Add 1/4 pound soaked boiled and crumbled saltfish.
Beat in 2 eggs.
Add a little baking powder, sauted onions, scallion, thyme.
Mix together.
Drop spoonfuls in hot oil.

—Alphanso McLean, Terra Nova Hotel, Kingston

Puerto Rico: Bacalaitos

Pupa is the popular nickname for Providencia Trabal, who is passionate about all Puerto Rican subjects. She used to demonstrate traditional Puerto Rican cooking on television. Now she cooks for relatives in the narrow high-ceilinged kitchen of her San Juan apartment. This is how she makes *Bacalaitos.*

About 2 cups wheat flour
1 or 2 spoonfuls of baking soda
Add to the last water from soaking the salt cod.
Work into a thick batter.
1/2 pound already boiled salt cod crumbled in
Add a spoonful of garlic chopped with oregano
Add 2 spoonfuls finely chopped onions
Add 2 spoonfuls finely chopped tomato
Add chopped coriander leaf and culantro *(local herb)*
Fry in hot corn oil dropping in a spoonful at a time from a ladle.

—Providencia Trabal, San Juan, 1996

"*Aye, Que Bonita!*" she exclaimed, and they are beautiful: two-inch amber puffs with the red and green of the herbs and vegetables brightened from quick cooking.

BRANDADE

Some believe *brandade de morue* began in Nîmes, but it is more commonly associated with Provence. It was originally called *branlade,* meaning "something that is pummeled," which it is. The dish had made it to Paris by the time of the French Revolution and never left. In 1894, writer Alphonse Daudet started a circle that met at the Café Voltaire on Place de l'Odéon for a regular *dîner de la brandade.*

Since salt cod has become expensive, potatoes have been added—*brandade de morue parmentier.* Antoine-Auguste Parmentier was an eighteenth-century officer who popularized the potato in the French Army, and his name has ever since meant "with potatoes." In 1886, *brandade* was decreed an official part of the enlisted man's mess in the French Army. As the price of salt cod has risen, so has the amount of potatoes in the *brandade.* Sometimes the dish simply seems like fishy mashed potatoes. As American Sara Josepha Hale wrote in her 1841 book, *The Good Housekeeper,* "The salted codfish is cheap food, if potatoes are used freely with it." The original *brandade* had no potatoes.

The following recipe, by the great nineteenth-century Provençal chef J.-B. Reboul, is especially flavorful because of the use of the skin.

MORUE EN BRANDADE

Use good salt cod, not too soaked and well scaled, cook as above (soaked 12 hours in fresh water, scaled and cut in squares. Put in a pot covered in cold water, put on the heat until a foam rises to the surface and skim it off), drain it. Carefully remove the bones, but leave the skin which contributes a great deal to the success of the operation. Put the well-cared-for pieces in a pot placed on a corner, so that it is gently heated with the milk in a small pot to one side and the oil in another, both moderately warm. Begin adding a spoonful of oil to the salt cod, work it strongly with a wooden spoon, crushing the piece against the sides of the pot, adding from time to time, little by little, the oil and the milk, alternating the two but always working hard with the wooden spoon. When the preparation becomes creamy, when you can no longer make out any pieces, the brandade *is finished.*

—J.-B. Reboul,
La Cuisinière Provençale, Marseille, 1910

The author goes on to suggest that truffles, lemon juice, white pepper, grated nutmeg, or garlic can be added and concludes by warning: "If we were health advisers, we might counsel you to use this dish in moderation."

THE FISH THAT SPOKE BASQUE

The most highly developed salt cod cuisine in the world is that of the Spanish Basque provinces. Until the nineteenth century, salt cod was exclusively food of the poor, usually broken up in stews. In PYSBE's 1936 collection of salt cod recipes, the largest section is devoted to stews. Few of these old-style salt cod dishes can still be found in the restaurants of the Basque provinces, but they are still made at home from the least expensive cut of *bacalao: desmigado* (trimmings). The most expensive cuts are tongues and *lomo*, the choice center cut of a fillet from high up near the head, cut from a larger cod.

With Cider

A salt cod omelette and *chuleta*—a shell steak, coated in salt and then grilled—are the two specialties of Basque cider mills. In both cases, the idea is to serve something salty to induce thirst. In San Sebastián's province of Guizpúzcoa, cider mills, *sidrerias*, are open only from January to April, during which time they try to lure as many people as possible to their tasting rooms so that they will have customers after the barrel-fermented cider is bottled in April. Customers are served food while standing at tall tables. Then, thirsty from the salt, they wander to the tasting room, sample, wander back and eat a little more, then taste some more. The cider room has barrels ten feet high. A hole is tapped, and customers stand in the middle of the room and catch the cider in large, straight-sided glasses, as it spouts from the hole. The glasses should be held vertically so that the cider hits the far side, not the bottom, creating a slight head as the taster walks his glass toward the barrel and then lifts it away, freeing the spout to land in the waiting glass of the next taster at the back of the room. Remarkably little cider ends up on the floor, which is probably proof of its low alcohol content.

The following recipe comes from a *sidreria* in a wooded mountain suburb of San Sebastián. The omelette has a wonderful salt cod taste, which is probably enhanced by using a far better cut than is traditional for this dish.

> *Soak the* lomo *for 36 hours and no longer to keep a little taste. Sauté chopped onions and a pinch of parsley in olive oil. Add the soaked and drained salt cod. Then add eggs beaten with a small amount of water. The secret is to do all this very quickly.*
>
> —Nati Sancho, cook for Sidreria Zelaia, 1996

Bacalao a la Vizcaína

In the nineteenth century, elegant salt cod dishes were created using a choice piece of *lomo,* always kept whole with the skin left on and served with a sauce. Three dishes became, and remain, dominant: *bacalao a la Vizcaína, al pil pil,* and *club ranero.* With their red, yellow, and orange sauces, the beauty of these dishes was part of their appeal. Like the standard repertoire of a concert violinist, all great Basque chefs must demonstrate some skill in these three dishes without taking liberties with the standard recipes. Great debates circulate over arcane issues such as the soaking of the fish. Should it be thirty-eight hours as Jenaro Pildain at Guria in Bilbao says, or forty-eight as recommended by Juan José Castillo at Casa Nicolas in San Sebastián? Pildain soaks it in the refrigerator. Castillo sometimes uses mineral water for soaking, claiming he detects a chlorine taste in the tap water.

Despite their elegance, these dishes used to appear in the most humble settings. Before the Spanish Civil war, a woman owned a tavern in Arakaldo, a small Basque village in Vizcaya. Typical of the inexpensive village eateries of the 1930s, the tavern in Arakaldo offered all the classic salt cod dishes to the poor people of the village. Her son worked with her and learned the repertoire. Today he is often referred to as *el rey de bacalao* (the king of salt cod). His famous restaurant on the main commercial street of Bilbao, Restaurante Guria, is considered the definitive place for the three classic dishes which he learned from his mother.

"Funny, it was food for poor people then. Now they are the most prestigious dishes I do," said Pildain.

Although he offered the following recipe, he also pointed out that it takes him a year to train a new cook to do the salt cod dishes.

An 1888 Spanish book made the claim that the two Spanish dishes most known in the rest of the world were *paella* and *bacalao a la Vizcaína.* More than 100 years later, this is still true. And yet *bacalao a la Vizcaína* is a dish that is almost impossible to reproduce. The sauce is based on a chubby little green pepper, the *choricero,* which grows to about three inches in length and then turns red and is dried. Until recently, the *choricero* grew only in the province of Vizcaya and is still native only to northern Spain.

In the Spanish-speaking Caribbean, where Cubans and Puerto Ricans regard this dish as part of their own national cuisine, their version does not even resemble the original. Not only is the pepper not available, but the West Indies quality of salt cod can only be broken up and stewed, usually with tomatoes and potatoes.

For 6 people:

12 pieces of salt cod 200 grams each
1 liter of vizcaína *sauce*
4 garlic cloves
1 liter of olive oil

Soak the salt cod for about 36 to 44 hours. During this time change the water every 8 hours. Taste to see if it has been long enough for the fish to be perfectly desalinated. If so, remove the salt cod from the water and drain it. Scale it well and remove bones.

Place a deep pan with oil and sliced garlic cloves on heat, remove the garlics once they are golden. Place the salt cod with the skin side up in the pan and poach for about 5 minutes. Remove the salt cod when well-cooked and pour on the vizcaína *sauce.*

For 1 liter of vizcaína *sauce:*

1 kilo of red and white onions
10 meaty choricero *peppers*
75 grams of ham
2 parsley bunches
½ liter olive oil
1 liter beef stock
30 grams of butter
3 garlic cloves
ground white pepper
salt

Put oil with garlic on heat in an aluminum pan. Once the garlic is golden add chopped onions, ham, and parsley, cooking strongly for 5 minutes and on low heat for another 30 minutes, stirring with a skimmer to avoid sticking to the pan. Open and remove seeds from the choricero *peppers and place in lukewarm water over heat. When it starts to boil add a little cold water to slow it down. Repeat this four times. Drain the peppers well and add to the already prepared mixture. Cook for 5 minutes over a low heat, take off the oil and the parsley and add the beef stock, the white pepper, and salt, letting it cook 15 minutes more. When well cooked, pass through a blender and then twice through a strainer. Put it back on the heat for 5 minutes, work in the butter, and adjust salt and pepper to make it perfect.*

—Jenaro Pildain, Restaurante Guria, Bilbao, 1996

HOW TO COOK THE LAST LARGE COD

ON CHOOSING A FRESH COD: "THE HEAD SHOULD BE LARGE; TAIL SMALL; SHOULDERS THICK; LIVER, CREAMY WHITE; AND THE SKIN CLEAR AND SILVERY WITH A BRONZE LIKE SHEEN."

—British Admiralty,
Manual of Naval Cookery, 1921

Only people who have lived by the North Atlantic understand the quality of fresh cod. It does not even resemble, except maybe in color, a fresh frozen cod. Fresh cod will inconveniently fall apart in cooking, which was why Sam Lee's New Orleans customer did not like his shipment. If it does not flake, it is not fresh. Fresh cod is "white, delicate, resilient," according to Paris chef Alain Senderens. "It will not tolerate long cooking. If you cook it carefully, cod will flake and give off milky cooking juices."

People who know fresh cod—from the great restaurants of France, to British working-class fish shops, to the St. John's waterfront—all agree on three things: It should be cooked quickly and gently, it should be prepared simply, and, above all, it must be a thick piece. Only a large piece can be properly cooked. The Lyons region's celebrated Paul Bocuse begins a simple recipe for fresh cod with potatoes and onions: "Use a piece of cod about 30 centimeters long cut from the center of the fish." The center of the fish is the thickest part. Bocuse is talking about the choice center of a three-foot cod, which is what everyone who knows fresh cod wants. But it is getting hard to find.

Alexandre Dumas gave these tips on selecting cod: "Choose a handsome spotted cod from Ostende or the Channel. . . . the best have white skin and yellow spots." He also offered the following recipe:

Breaded Cod

Cut the cod in five or six pieces, marinate with salt, pepper, parsley, shallots, garlic, thyme, bay leaf, green onions, basil; all chopped, the juice of two lemons and melted butter, prepare it with the marinade and bread it and cook it in a country oven.

—Alexandre Dumas,
Le Grande Dictionnaire de cuisine, 1873 (posthumous)

Cod Bonding

Wherever there are Norwegian communities, there are cod clubs. There is one in New York and four in the Minneapolis–

St. Paul area. The clubs are usually exclusively for men. According to Bjarne Grindem, the former Norwegian consul in Minneapolis, three of the four in the Twin Cities are all men, and the fourth is "more liberal." Although cod clubs claim to be exclusive and applicants wait for years for a place, each club has as many as 200 members. One hundred or more men get together once a month at lunchtime, and the meal is always boiled cod and potatoes with melted butter served with aquavit and flat bread, called *kavli.* "Whether they get together to get together or get together to eat cod is another question, but they always get together around the cod," said Grindem. The oldest, most exclusive of the Twin City clubs is the Norwegian Codfish Club at the Interlochen country club in Edina. While the members gave lectures on the exact way to prepare a boiled cod, never letting the water actually boil, the kitchen at the Interlochen was more prosaic: "You mix salt water and bring to a boil and put the fish in and cover and cook for half an hour. It's a good thick fish, about a pound a person."

The Last of the Northern Stock

Stella's is a popular, cozy little restaurant on the St. John's waterfront. Miraculously, one day the restaurant was able to buy enough large, thick, cod fillets from the Sentinel Fishery to put this old standard back on the menu for one night—just a teaser, reminding Newfoundlanders of what they were missing. Stella's defies Newfoundland tradition and refuses to use pork fat, understandably regarding it as unhealthy.

Panfried Cod

4 fresh cod fillets
2 eggs beaten with 1/4 cup milk
1 cup flour mixed with 1 teaspoon paprika, 1/4 teaspoon black pepper, 1 teaspoon parsley

Dip fish in egg mixture, then in flour mixture. Have pan hot. Then fry in vegetable oil—very hot, then as soon as you put the fish in, turn it down.

—Mary Thornhill, Stella's Restaurant, St. John's, 1996

MEASUREMENT EQUIVALENTS

1 gram	=	0.35 ounces
1 kilogram	=	2.2 pounds
1 milliliter	=	.03 fluid ounces
1 deciliter	=	.3 fluid ounces
1 liter	=	2.1 pints
		or 1.06 quarts
		or .26 gallons
1 teaspoon	=	5 milliliters
1 tablespoon	=	15 milliliters
1 cup	=	.24 liters
1 pint	=	.47 liters
1 quart	=	.95 liters

Bibliography

GENERAL HISTORY

Auden, W. H., and Louis MacNeice. *Letters from Iceland.* London: Faber and Faber, 1967.

Babson, John J. *History of the Town of Gloucester, Cape Ann.* Introduction by Joseph E. Garland. Gloucester: Peter Smith, 1972.

Boorstin, Daniel J. *The Americans.* 3 vols. Vol. 1, *The Colonial Experience.* Vol. 2, *The National Experience.* Vol. 3, *The Democratic Experience. New York:* Random House, 1958, 1965, 1973.

Davant, Jean-Louis. *Histoire du Peuple Basque.* Bayonne: Elkor, 1996.

Draper, Theodore. *A Struggle for Power: The American Revolution.* New York: Times, 1996.

Fabre-Vassas, Claudine. *The Singular Beast: Jews, Christians and the Pig,* trans. Carol Volk. New York: Columbia University Press, 1997.

Felt, Joseph B. *Annals of Salem.* Vol. 1 Boston: W. and S. B. Ives, 1945.

Huxley, Thomas Henry. *Man's Place in Nature and Other Essays.* London: J. M. Dent and Sons, 1906.

Lukas, J. Anthony. *Common Ground: A Turbulent Decade in the Lives of Three American Families.* New York: Knopf, 1985.

Massachusetts House of Representatives, compiled by a committee of the House. *A History of the Emblem of the Codfish in the Hall of the House of Representatives.* Boston: Wright and Potter Printing, 1895.

Miller, William Lee. *Arguing About Slavery: The Great Battle in the United States Congress.* New York, Knopf, 1995.

Morison, Samuel Eliot. *The Great Explorers: The European Discovery of America.* New York: Oxford University Press, 1978.

Nugent, Maria. *Lady Nugent's Journal: Jamaica One Hundred and Thirty Years Ago.* Edited by Frank Cundall. London: West India Committee, 1934.

Perley, Sidney. *The History of Salem Massachusetts.* Salem: published by the author, 1924.

Smith, Adam. *An Inquiry into the Nature and Causes of the Wealth of Nations.* Edited by Edwin Cannon. New York: Modern Library, 1937. First published in 1776.

Thoreau, Henry David. *Cape Cod.* New York: Penguin, 1987. First published in 1865.

FISH AND FISHERIES

Binkley, Marian. *Voices from Off Shore: Narratives of Risk and Danger in the Nova Scotian Deep-Sea Fishery.* St. John's: Iser, 1994.

Blades, Kent. *Net Destruction: The Death of Atlantic Canada's Fisheries.* Halifax: Nimbus, 1995.

Butler, James Davie. "Codfish; Its Place in American History." Transactions of the Wisconsin Academy of Science, Arts, and Letters, vol. 11, 1897.

Chantraine, Pol. *The Last Cod Fish: Life and Death of the Newfoundland Way of Life.* Toronto: Robert Davies, 1994.

Clement, Wallace. *The Struggle to Organize: Resistance in Canada's Fishery.* Toronto: McClelland and Stewart, 1986.

Collins, Captain J. W. *Howard Blackburn's Fearful Experience of a Gloucester Halibut Fisherman, Astray in a Dory in a*

Gale off the Newfoundland Coast in Midwinter. Gloucester: Ten Pound Island, 1987.

Convenant, René. *Galériens des Brumes: Sur les voiliers Terre-neuvas.* St.-Malo: L'Ancre de Marine, 1988.

Doel, Priscilla. *Port O'Call: Memories of the Portuguese White Fleet in St. John's, Newfoundland.* St. John's: Iser, 1992.

Earle, Liz. *Cod Liver Oil.* London: Boxtree, 1995.

Garland, Joseph E. *Gloucester on the Wind: America's Greatest Fishing Port in the Days of Sail.* Dover, N.H.: Arcadia, 1995.

Grzimek's *Animal Life Encyclopedia.* Vol. 4, *Fishes I.* New York: Van Nostrand Reinhold, 1973.

Homans, J. Smith, and J. Smith Homans, Jr., ed. *Cyclopedia of Commerce and Commercial Navigation.* New York: Harper and Brothers, 1858.

Innis, Harold A. *The Cod Fisheries: The History of an International Economy.* New Haven: Yale University Press, 1940.

Jentoft, Svein. *Dangling Lines: The Fisheries Crisis and the Future of Coastal Communities: The Norwegian Experience.* St. John's: Iser, 1993.

Joncas, L. Z. *The Fisheries of Canada.* Ottawa: Office of the Ministry of Agriculture, 1885.

Jónsson, J. "Fisheries off Iceland, 1600–1900." *ICES Marine Science Symposium,* 198 (1994): 3–16.

Kipling, Rudyard. *Captains Courageous.* Pleasantville, N.Y.: Reader's Digest, 1994.

Martin, Cabot. *No Fish and Our Lives: Some Survival Notes for Newfoundland.* St. John's: Creative Publishers, 1992.

McCloskey, William. *Fish Decks: Seafarers of the North Atlantic.* New York: Paragon, 1990.

Melville, Herman. *Moby Dick or The Whale.* New York: Random House, 1930. First published in 1851.

Ministry of Agriculture and Fisheries. *Fishery Investigations,* series 2, vol. 7, no. 7. London, 1923.

Montgomery, Doris. *The Gaspé Coast in Focus.* New York: E. P. Dutton, 1940.

Parsons, Gillian A. *The Influence of Thomas Henry Huxley on the Nineteenth-Century English Sea Fisheries.* Lancaster: University of Lancaster, 1994.

———. "Property, Profit, Politics, and Pollution: Conflicts in Estuarine Fisheries Management, 1800–1915." Doctoral dissertation, University of Lancaster, 1996.

Pierce, Wesley George. *Goin' Fishin': The Story of the Deep-Sea Fishermen of New England.* Salem: Marine Research Society, 1934.

Ryan, Shannon. *Fish out of Water: The Newfoundland Saltfish Trade,* 1814–1914. St. John's: Breakwater, 1986.

Storey, Norman. *What Price Cod? A Tugmaster's View of the Cod Wars.* North Humberside: Hutton Press, 1992.

Taggart, C. T.; J. Anderson; C. Bishop; E. Colbourne; J. Hutchings; G. Lilly; J. Morgan; E. Murphy; R. Myers; G. Rose; and P. Shelton. "Overview of Cod Stocks, Biology, and Environment in the Northwest Atlantic Region of Newfoundland, with Emphasis on Northern Cod." *ICES Marine Science Symposium* 198 (1994): 140–57.

Thór, Jón Th. *British Trawlers and Iceland, 1919–1976.* Göteborg, Sweden: University of Göteborg, 1995.

Yvon, R. P. *Avec les pecheurs de Terre-Neuve et du Groenland.* St.-Malo: L'Ancre de Marine, 1993.

FOOD HISTORY AND ANTHROPOLOGY

Allen, Brigid. *Food: An Oxford Anthology.* Oxford: Oxford University, 1994.

Artusi, Pellegrino. *The Art of Eating Well.* Translated by Kyle M. Phillips III. New York: Random House, 1996.

Brereton, Georgina E., and Janet M. Ferrier, ed. *Le Mesnagier de Paris.* Paris: Le Livre de Poche, 1994.

Brillat-Savarin, Jean Anthelme. *Physiologie du goût.* Paris: Flammarion, 1982. First published in 1825.

Capel, José Carlos. *Manual del pescado.* San Sebastián: R and B, 1995.

Collins, John. *Salt and Fishery, Discourse Thereof.* London, 1682.

Couderc, Philippe. *Les Plats qui ont fait La France: De l'andouillette au vol-au-vent.* Paris: Julliard, 1995.

Davidson, Alan. *North Atlantic Seafood.* London: Macmillan, 1979.

De Andrade, Margarette. *Brazilian Cookery: Traditional and Modern.* Rio de Janeiro: A Casa do Livro Eldorado, 1985.

de la Falaise, Maxime. *Seven Centuries of English Cooking: A Collection of Recipes.* Edited by Arabella Boxer. New York: Grove Press, 1973.

Dumas, Alexandre. *Le Grand Dictionnaire de cuisine.* Vol. 3, *Poissons.* Payré: Édit-France, 1995. First published in 1873.

Hieatt, Constance B., ed. *An Ordinance of Pottage: An Edition of the Fifteenth Century Culinary Recipes in Yale University's Ms Beinecke 163.* London: Prospect, 1988.

Hope, Annette. *A Caledonian Feast.* London: Grafton, 1989.

McClane, A. J. *The Encyclopedia of Fish Cookery.* New York: Henry Holt, 1977.

McGee, Harold. *On Food and Cooking: The Science and Lore of the Kitchen.* New York: Scribner's, 1984.

Montagne, Prosper. *Larousse gastronomique.* Paris: Larousse, 1938.

Oliver, Sandra L. *Saltwater Foodways: New Englanders and Their Food, at Sea and Ashore, in the Nineteenth Century.* Mystic, Conn.: Mystic Seaport Museum, 1995.

Otaegui, Carmen. *El Bacalao: Joya de nuestra cocina.* Bilbao: BBK, 1993.

Pizer, Vernon. *Eat the Grapes Downward: An Uninhibited Romp through the Surprising World of Food.* New York: Dodd, Mead, 1983.

Riley, Gillian. *Renaissance Recipes.* London: New England Editions, 1993.

Root, Waverley. *Food.* New York: Simon and Schuster, 1980.

Root, Waverley, and Richard de Rochemont. *Eating in America: A History.* New York: William Morrow, 1976.

Scully, Terence, ed. *The Viandier of Taillevent.* Ottawa: University of Ottawa Press, 1988.

Sloat, Caroline, ed. *Old Sturbridge Village Cookbook.* Old Saybrook, Conn.: Globe Pequot Press, 1984.

Tannahill, Reay. *Food in History.* New York: Stein and Day, 1973.

Thorne, John, with Matt Lewis Thorne. *Serious Pig: An American Cook in Search of His Roots.* New York: North Point Press, 1996.

Toussaint-Samat, Maguelonne. *History of Food.* Translated by Anthea Bell. Oxford: Blackwell, 1992.

Trager, James. *The Food Chronology: A Food Lover's Compendium of Events and Anecdotes, from Prehistory to the Present.* New York: Henry Holt, 1995.

Visser, Margaret. *Much Depends on Dinner: The Extraordinary History and Mythology, Allure and Obsessions, Perils and Taboos of an Ordinary Meal.* New York: Macmillan, 1986.

Wilson, C. Anne. *Food and Drink in Britain: From the Stone Age to the 19th Century.* Chicago: Academy Publishers, 1991.

Acknowledgments

This book owes a great debt to many people throughout the Atlantic region, but especially: José Juan Castillo in San Sebastián, Hallfredur Örn Eiríksson at the Árni Magnússon Institute in Reykjavík, Lillian Gonzalez in New York, Einar Gustavsson at the Scandinavian Tourist Board in New York, Jørgen Leth in Port-au-Prince, Louis Menashe in New York, David S. Miller in Copenhagen, Rosita Marrero in San Juan, Gillian Parsons at the University of Lancaster, Christine Toomey in London, John Walton at the University of Lancaster, and Monique Zerdoun in Paris. Special thanks to Lisa Klausner.

I am greatly indebted to Denise Martin and Linda Perney for so generously giving their time and considerable skills and for much valued advice. I also want to thank Charlotte Sheedy for all her support and guidance. This book owes much to my incredibly good fortune in finding the right editor and publisher. Deep-felt thanks to Nancy Miller for her friendship, faith, enthusiasm, and skillful work and to George Gibson, who is the kind of publisher of which most writers only dream.

Index